# ENGINEERING ◆ NATURE

The University of North Carolina Press    Chapel Hill

JESSICA B. TEISCH

# ENGINEERING ◊ NATURE

## Water, Development, & the Global Spread of
## American Environmental Expertise

This book was published with the assistance of the Anniversary
Endowment Fund of the University of North Carolina Press.

Designed by Courtney Leigh Baker and set in Merlo, Gotham, and
Clarendon by Tseng Information Systems, Inc. Parts of this book have
been reprinted with permission in revised form from "Sugar Economies
in the Far West: California's Hawaiian Outpost," in *Cities and Nature in the
American West*, edited by Char Miller (Reno: University of Nevada Press, 2010),
and "'Home Is Not So Very Far Away': Californian Engineers in South Africa,
1868–1915," *Australian Economic History Review* 45, no. 2 (2005): 139–60.
The paper in this book meets the guidelines for permanence and durability of the Committee on Production Guidelines for Book Longevity
of the Council on Library Resources. The University of North Carolina
Press has been a member of the Green Press Initiative since 2003.

Library of Congress Cataloging-in-Publication Data
Teisch, Jessica B.
Engineering nature : water, development, and the global spread of
American environmental expertise / Jessica B. Teisch
p. cm.
Includes bibliographical references and index.
ISBN 978-0-8078-3443-5 (cloth : alk. paper)
ISBN 978-0-8078-7176-8 (pbk : alk. paper)
1. Water resources development—United States—History. 2. Mining
engineering—United States—History. 3. Water resources development—
History. 4. Mining engineering—History. I. Title.
HD1694.A5T37 2011
333.70973—dc22   2010029268

cloth   14 13 12 11 10   5 4 3 2 1
paper   14 13 12 11 10   5 4 3 2 1

THIS BOOK WAS DIGITALLY PRINTED.

To CAROLYN MERCHANT for her
intellectual generosity and encouragement and
MICHAEL JOHNS for his unflagging support,
and in memory of my friend HAL K. ROTHMAN,
who remains a source of inspiration

# CONTENTS

# ILLUSTRATIONS

ACKNOWLEDGMENTS

I first encountered what was then the burgeoning field of environmental history in 1993, while enrolled as an undergraduate in a course taught by Caroline Karp at Brown University. *Engineering Nature* is, in some ways, an outgrowth of the questions about the relationship between people and the natural world I grappled with at that time. Although the field has become decidedly more complex and this book shares a broader scope with many disciplines, the fundamental economic, social, technical, and cultural questions about our relationship to our environment persist.

Among the many people I wish to thank for making this book possible, Carolyn Merchant and Michael Johns are the first. Much appreciation also goes to Carl Abbott, Marc Cioc, David Igler, Char Miller, Donald Pisani, Charles Postel, Mary Ryan, Michael Watts, Wendy Wolford, the dedicated staff at the Bancroft Library at the University of California at Berkeley, Mark Simpson-Vos and the editors at the University of North Carolina Press, and my wonderful family—my parents, Joel and Fran; my sisters, Rachel and Abby; and my husband, Michael, and our son, Jonah. Finally, this book—and the travels on which I embarked while writing it—would not have been possible without the generosity of the Jacob K. Javits Fellowship from the U.S. Department of Education and the Bancroft Library Research Award.

# California Welcomes the World

All of the great world's fairs of history—London, Philadelphia, Chicago, St. Louis, Paris—gazed back in time and recorded the progress made by important events, discoveries, and inventions. They also looked forward. The first international fair in 1851 portrayed a world dominated by the new industrial capitalism. Exhibits in London's Hyde Park represented an ordered world where people bought and sold goods and exchanged ideas about art, science, technology, education, and the relations among nations. The magical world's fairs that followed celebrated the march of human progress by honoring these themes.[1]

San Francisco's Panama-Pacific International Exposition of 1915 was no different. The state's nineteenth-century dream had rested on anti-industrial and agrarian themes, but the 1915 exposition announced California's future. In the Palace of Transportation, a Ford rolled off an assembly line every ten minutes. The Palace of Machinery displayed life-sized models of California's mechanized canneries, cement mixers, electricity-lit mines, Pelton water wheels, power plants, and diesel engines. The fair also gave one last tribute to the ideals of a century that had been defined, above all, by one of the most self-confident and exuberant ideas of all times: progress.[2]

This nineteenth- and early-twentieth-century notion of progress rested on an abstract but universal set of ideas. In theory, progress promised many things: material and commercial development, scientific and social enlightenment, free markets, and rule of law. The San Francisco Exposition's machine exhibits, international pavilions, and educational displays honored these ideals. Most of all, the fair celebrated technical innovation.

Paris had the Eiffel Tower; Chicago had its Ferris wheel; San Francisco had the Panama Canal. Perhaps nothing symbolized the modern era more than this canal, the world's largest engineering project to date. Nations had dreamed of constructing a waterway through the Isthmus of Panama since Balboa landed there in 1500. This vision spilled over into the next four centuries. In the mid- to late nineteenth century, applied science in the forms of civil engineering, fluid mechanics, and hydraulics made large engineering works possible. In the 1890s, international teams of engineers and workers began to use dredges, steam shovels, cranes, locomotives, and drills to cut a canal through fifty miles of Panama's hills and rocks. By 1914, they had excavated a mass of earth larger than the Egyptian pyramids and built the world's greatest water bridge. The fair's two million visitors walked through the San Francisco Exposition's rose-colored gates just as the Panama Canal admitted the first ships through its locks.[3]

The fair's working model of the canal had "oceans" at each end and a moving platform with 144 miniature double-decker electrical cars that transported visitors from the "Pacific" to the "Atlantic" in twenty-three minutes. Spectators sat in comfortable chairs and held telephone receivers to their ears. Phonograph records, which operators switched on and off to correspond with the various sections of the journey, relayed recorded lectures on the canal. Visitors learned about how the locks opened, closed, and admitted vessels. Fairgoers marveled at models of Panama's rivers, mountains, existing and submerged cities, dams, lighthouses, "electric mules," and power plants.[4] The model canal was an exercise in geography, engineering, and culture, a lesson that stamped onto visitors' minds the crowning achievement of a half century: the conquest of technology over the natural environment and, as the *San Francisco Examiner* remarked, humankind's "mastery of this planet."[5]

In the mid–nineteenth century, modern transportation and communication systems helped set our modern world in motion. The application of iron, steel, and fossil fuels to industry, transportation, and communication changed the way that people and nations interacted. First, these changes accelerated growth in material production. Second, they helped integrate a global economy. The Suez Canal, steamships, submarine telegraph cables, and transcontinental railways locked distant places together in modern financial and commodity webs. Karl Marx and Friedrich Engels noted in 1848 that capitalist production "must nestle everywhere, settle everywhere, establish connexions everywhere." Capitalism "draws all, even the most barbarian, nations into civilization [and] creates a world after its own image."[6] The rise of modern capitalism, which marked the beginning of a half century of economic liberalism, touched almost every part of the world. For the first time, developments local to one region spread quickly to others.[7]

The third effect of modern technology, a direct consequence of the sec-

*Construction of the Panama Canal's Miraflores Upper Locks, July 13, 1910.*
*Courtesy of the Bancroft Library, University of California, Berkeley*

*"Panorama Canal," the Panama Canal exhibit at San Francisco's Panama-Pacific International Exposition of 1915. Courtesy of the Bancroft Library, University of California, Berkeley*

ond, was the repartitioning of the world into unequal parts. That powerful fusion of machines, markets, and money was most obvious in Western Europe and the United States, where technical innovation was chiefly indigenous. By the 1880s, more than one hundred thousand railway locomotives and twenty-two thousand steamships crossed the continents and oceans. People from these regions made two billion trips by rail annually. In other, much larger areas, including Africa, Asia, and South America, progress came as a foreign conqueror.[8]

Part of the economic and military supremacy of the Western capitalist nations involved the conquest, annexation, and administration of non-Western countries. Between 1880 and 1914, much of the world outside of Europe and the Americas was divided into territories under the formal rule or informal political domination of one of a handful of states: Great Britain, France, Germany, Italy, Portugal, Russia, Italy, the Netherlands, Belgium, the United States, and Japan. European powers almost entirely carved up two major regions of the world, Africa and the Pacific.

The spreading global economy generated vast inequalities. It was a time, after all, that Eric Hobsbawm calls the Age of Imperialism, a time that brought places together while simultaneously generating regional inequalities and divergent kinds of modernities. The era that in many ways brought disparate regions together with technology and world markets thus also found new ways of keeping them apart.[9]

Despite growing national hierarchies, people believed more in technology's transformative powers at the end of the nineteenth century than perhaps at any other time. At an 1895 meeting of the American Society of Civil Engineers, prominent American engineer George S. Morison, who later joined the Isthmian Canal Commission (the American bureaucracy that oversaw the early years of construction of the Panama Canal), envisioned machines, railways, and dams leveling regional differences in the productive abilities of states and nations. Morison had achieved recognition by applying steel to bridge and railroad design throughout America's Northeast and Midwest. Witnessing his success in facilitating transportation over the Missouri, Columbia, Snake, and Mississippi Rivers, he optimistically predicted that modern machinery, combined with new sources of power, would draw a line between the "ancient and the modern, between ignorance and intelligence, between . . . barbarism, and the new civilization." He foresaw that human intellect and capacity, which guided technical progress, would overcome the environmental and racial obstacles that had constrained the progress of previous centuries.[10] Technology also gave people the "means of traversing the entire globe with a regularity and speed which brings all races together [and will] finally make the human race a single great whole working intelligently in ways and for ends which we cannot yet understand." He concluded that humankind stood "at the dawn of a great epoch, of an epoch making greater changes than the world has ever seen."[11]

If we now view Morison's claims with skepticism, it is nonetheless easy to see how he reached those conclusions. The Panama Canal embodied Morison's belief that technology drove the markets that were spreading throughout the world. According to San Francisco Exposition president Charles C. Moore, the first professional engineer to organize and host a world's fair, the waterway encouraged U.S. "acquaintance with nations destined to become increasing factors" in commerce.[12] The canal provided a thoroughfare that enabled ocean steamers to make swift runs between San Francisco, London, Liverpool, New York, Yokohama, Hong Kong, and Manila. It tied together the West and East Coasts of the Americas.[13] "Inherently," a writer for *Sunset Magazine* noted, "the Panama-Pacific Exposition is a celebration of the lessening of distance in one world-instance."[14]

This shrinking world meant uniting distant cultures, an idea represented at the exposition. In the fair's Court of the Universe, a statue of a serene Asian gentleman greeted an ox-drawn prairie schooner filled with American Western pioneers. One visitor noted that this symbolic encounter between East and West typified "the meeting of the world's families now that the Canal has been completed."[15] A fifty-million-dollar Palace of World Peace stood "as a glorious embodiment of the dream of a federated world," despite the outbreak of war in Europe the preceding August.[16]

Ironically, an attitude of conquest drove the exposition's vision of global unity. The Panama Canal, one of the world's greatest technical feats, also represented the utopian belief in the need for an advanced social order demonstrating humanity's progress. Uniting the world and creating a new social order entailed conquering many parts of the non-English-speaking world; progress and racial subjugation went hand in hand.

This progressive view of history emerged in the last half of the nineteenth century, when scientists extended the study of biology to the subject of race. The phrase "Ascent of Man," an idea that extended well past the Victorian era, told the story of humankind.[17] Why should the same laws that characterized the natural world, which Charles Darwin had shown to be the product of natural selection, not also govern society? According to the predominant thinking, genetic and environmental inequalities had doomed some people to inferiority and had marked other people—those of Anglo-Saxon and European stock—for superiority and for carrying civilization to the dark corners of the earth.[18]

The fair exhibited the theme of Social Darwinism and its sister science, eugenics, in great splendor.[19] The Court of the Universe displayed forty-foot-high scenes from "Nations of the East" and "Nations of the West." The former showed black slaves carrying baskets of fruit on their heads and depicted Mongolian warriors as "the ancient inhabitant of the sandy waste." By contrast, the Western panels idealized Western progress with a westward-facing statue of the Heroes of Tomorrow—a prairie schooner with an Italian immigrant, Anglo-American, squaw and papoose, and Plains Indian in tow.[20] National pavilions confirmed these stereotypes. Bolivia erected a crude stone building expressing the presumed simplicity of its Spanish and Indian populations. Guatemala, Honduras, and Panama displayed the products of their forests, fields, and mines. They had not yet, according to one critic, developed "characteristic styles of their own."[21] France, by contrast, constructed an impressive palace that housed paintings from the Louvre, many illustrating Europe's glorious rise from barbarism.[22]

The Panama-Pacific Exposition even resulted in the formation of the World's Social Progress Council, which met for the first time in San Francisco. Led by Californians, the council recognized that some environments negatively affected various cultures and called for the further development of the inferior races.[23] One writer for *Sunset Magazine* captured the nature of this progress: "America found Panama a dirty barbaric child sitting in her filthy gutters making mud pies. She scrubbed, scoured and combed her out. [The] threat of spanking has triumphed over natural inclination."[24]

If the fair presented a crude portrait of the less developed countries, it also pointed to the growing reality of America's mounting world influence and dominance. As the nineteenth century drew to a close, the United States became one of the leading import-export centers for the discovery and cross-fertilization of knowledge and techniques. American exports played increasingly significant roles in national economies. According to historian Daniel Rodgers, nations borrowed, copied, and adapted from others from the 1870s onward. Merle Curti and Kendall Birr further show that American exports played important roles in the late nineteenth century and continued to do so in the twentieth. Many of these exchanges were scientific and technical in nature. In 1874, for example, American geologists opened up black diamond districts in Brazil. The next decade, the U.S. Coast and Geodetic Survey led an expedition to Africa's west coast. In 1886, the Chinese government hired American engineers to introduce new mining methods. Shortly thereafter, the Pan-American Conference commissioned American engineers to chart a continental railway through Central and South America. In 1911, the American Red Cross sent a civil engineer to China to study river conservation. In 1916, the Panamanian government sent a commission to Washington, D.C., to elicit cooperation from highway, railway, radio, and agricultural developers. And in 1919, Herbert Hoover, the most famous mining engineer of his time, directed the American Relief Administration in Russia and Western Europe.[25]

Although transfers of American expertise increased over the course of the late 1800s and early 1900s, many of these exports, Rodgers argues, were not homegrown. Americans generally saw themselves as isolationist in the first half of the twentieth century, but they were in fact engaged with innovations in social policy, industry, art, and technology from Western Europe. German engineers came to the United States in the 1920s to study American factories; Irish, Austrian, and German agronomists investigated the rural cooperatives of Denmark; and British reformers reported on German leader Otto von Bismarck's social insurance schemes. "Tap into the debates that swirled through the United States and industrialized Europe over the problems and miseries of 'great city' life, the insecurities of wage work, the social backwardness of the countryside, or the instabilities of the market itself," Rodgers writes, "and one finds oneself pulled into an intense, transnational traffic in reform ideas, policies, and legislative devices."[26]

As this global "traffic" suggests, many of these exchanges were not only technical but also ideological and social. Prominent among the international group of travelers were Americans seeking lessons in social reform. Jane Addams, for example, visited London's Toynbee Hall in the 1880s. Richard T. Ely and W. E. B. Du Bois studied the social doctrines of the German Kathedersozialisten. And in the 1930s, Lewis Mumford and Catherine Bauer learned from modernist architecture and housing reform in Frankfurt, Berlin, and Vienna. Americans' notions of workers' insurance, subsidized housing, city planning, and rural reconstruction similarly derived from abroad. Most significantly, according to Rodgers, parts of America's New Deal were adopted from a transatlantic marketplace of progressive ideas that had been brewing since the 1880s.[27]

National and regional contexts, however, dictated how these different plans came together. In the course of exchanging information, goods, and services, these world travelers "assembled, dismantled, and reassembled" foreign cultures, as anthropologist Eric Wolf argues, "conveying in variable accents the divergent paths of groups and classes." Our twenty-first-century perspective allows us to understand that the "manner of that mobilization sets the terms of history, and in these terms the peoples who have asserted a privileged relation with history and the peoples to whom history has been denied encounter a common destiny."[28]

As evidenced by these international transfers of expertise on full display at the Panama-Pacific International Exposition, engineers were prominent in building this common world destiny. Although this common world cultured and advanced transnational technologies and global markets, it also fostered regional hierarchies and justified ideas of Social Darwinism, racial subjugation, manifest destiny, and Anglo-Saxon and European supremacy. The exposition's powerful rhetoric about the conquest of nature and cultures revealed that engineering was a civilizing task at heart, a project whose logic belonged to the nineteenth century's lofty dream of progress.

Frederick H. Newell, a prominent civil engineer and the first director of the U.S. Reclamation Service, called this era of rapid change the Age of the Engineer.[29] In many respects, the profession represented the age's perfect faith in progress. Engineers designed railroads through mountains, wired continents together in elaborate communication networks, excavated gold from deep inside the earth, and siphoned water from faraway rivers to deserts. Nor should their technological utopianism be underemphasized. These men (it was almost exclusively a male profession) believed that technical development would help settle or repopulate barren areas, create markets, and help further regions' entries into the global economy. Engineers thought that technology would bring myriad social benefits as well. It is not surprising that many engineers, according to Newell, viewed themselves as "missionar[ies] of light and progress" and pioneers of a "better and higher degree of civilization."[30] As professional men, engineers were also committed to applying scientific principles to business and society and acting as arbiters of social and class conflicts.[31] Thus, while engineers used their technology to alter nature, they also tried to redesign elements of society.

As the idea of universal progress took on different permutations around the world, the Age of the Engineer presented unparalleled opportunities for itinerant engineers, who traded variations of this dream.[32] Expertise that began in one region moved to others via modern transportation and communication systems. When Australian engineers met Americans on Indian soil and American engineers encountered British engineers in South Africa's gold fields, men exchanged ideas, technology, and experience. A talented river engineer who started his career on the Danube might move to the Mississippi and then reappear in the Amazon, using the same template to reengineer these rivers. Dam technology invented in Switzerland might migrate to South Africa. Although the host country supplied a wealth of new information, engineers crisscrossed the globe, often carrying similar models for remaking the natural landscape. By the late nineteenth century, knowledge about engineering as well as questions about technology's proper political and social context circulated the world. Engineers believed they had so corrected the "wrongs of nature" that John Hays Hammond, one of America's preeminent mining engineers, remarked, "Those portions of the globe which for eons had remained comparatively barren and useless now are being transformed into a blessing to man."[33]

One subset of this scientific and technical cadre, California engineers, was especially influential in shaping particular types of development throughout the world. They formed part of a worldwide epistemic community that linked together countries and peoples thousands of miles apart. California engineers did not create their global technical community per se; rather, this international fraternity of experts emerged from three centuries of European innovation that started in Italy, moved to France and Holland, and then spread to North America and beyond. But by the early twentieth century, California engineers were certainly among the most renowned water technicians in the world—perhaps second only to the Ruhr engineers—and exerted an enormous impact, especially in the Pacific Rim. Together, California's engineers and their tools, ideas, and models of progress offer one of the best illustrations of the uneven outcomes of the global application of new technologies and their motivating ideas.

Engineers' mining and irrigation projects in parts of California, Australia, South Africa, Hawaii, and Palestine show how Californians' transfers aided the entry of frontier regions into a global economy. The work of these engineers also suggests that globalization and the expansion of world markets and communication and transportation systems resulted from highly personal philosophies and endeavors. For California engineers, progress meant building major infrastructural projects. It also meant remaking the world in ways that affirmed their individual politics and beliefs as well as those of their society. Yet each country's historical context and institutional arrangements dictated how the Californians' projects came together. These variable responses, in turn, show the complex process of globalization at work between the 1870s and 1920s. How California engineers used technology as the vehicle to drive global political, ideological, social, economic, and environmental change shows the geographically distinct variations that "progress" created.

In the 1850s, California became a hub for the global spread of mining and irrigation technology as well as ideas about labor, property rights, government, society, and modern rural life. The gold rush gave engineers unparalleled technical expertise. Over the next decades, they fanned out across the globe with blueprints for change. Some of the tools and ideas that engineers transferred were indigenous to California. Others evolved from experiences learned abroad, were modified to fit California's conditions, and then were reexported to other countries.

Engineers saw many parallels between nineteenth-century California, Australia, South Africa, Hawaii, and Palestine. In each region, white settlers, operating under imperialist ideals, seized land. Local cultures and governments underwent shifts and upheavals. Settlers forged new social and racial hierarchies and replaced existing legal structures (and native peoples) with what they believed were more enlightened Western ways. As the scale of agriculture and industry increased to fit the needs of growing populations and economies, these regions struggled with similar issues regarding the relative roles of the market, state, property, indigenous peoples, and labor in a growing economy.

Californians exported a diverse set of ideas, methods, and tools. Individual components went to various places at different times. Taken as a whole, this bundle provided host countries with American models of development. Technology was the crucial item. California's engineers transferred blueprints of dams, canals, and mines. They used their knowledge of how to separate gold from ore and generate electricity from falling water. They also transported steel pipes, turbine engines, cement mixers, and waterwheels. Many of the regions in which California engineers worked resembled their home state's semiarid climate, topography, and geology. Although Australia's rivers were shallower than California's, South Africa's gold buried deeper, Hawaii's clime more humid, and Palestine's soils drier, California's tools and engineering plans generally fit into the overall scheme of development.

Transferring technology to foreign countries required vast knowledge about the natural world. During the Civil War, federal agencies had begun to map the American West's topography, minerals, rivers, watersheds, and possible railroad and canal routes. Government surveys, many of them headed by California engineers and scientists, devised new ways of measuring rainfall, stream flow, and soil type. Engineers subsequently carried their surveying instruments and scientific methods abroad.

Californians also tried to export the contextual frameworks for their scientific knowledge and tools. The hydraulic artillery that miners used to blast gold from the Sierra Nevada's mountainsides, for example, evolved out of a legal framework that allocated water to the first claimant. Loose government regulations gave rise to California's monopolistic land and water practices, profligate resource users, and battles between farmers and miners. Measures designed to address these conflicts included the creation of small irrigation colonies, local irrigation districts, new water laws, and expert-driven government bureaucracies. Engineers' tool kits contained not only technical and scientific knowledge but also ideas about frontier development, social and resource institutions, capitalism, government, and modern rural life.

California's experiences and personalities shaped infrastructural development and progress in frontier societies. Yet in practice, engineers' exchanges never played out as intended. The transfer of expertise requires the physical relocation of experts and technology. It also entails, at least in theory, the diffusion from one society to another of the technology's context.[34] The key to applying a body of knowledge was understanding its framework at home and adapting that context to conditions abroad. But what flourished in some regions floundered in others. Controlling the vagaries of nature required more than blocking rivers with dams. It also involved, for example, reconciling customary rights with new laws and political structures. Despite similar geographies, universal models of development frequently conflicted with existing and historical social, political, economic, and legal administrations.

While Californians focused on similarities among regions they visited, they tended to overlook differences. Hammond, for example, viewed the engineer as the "most important missionary of civilization." As a prominent member of this elite group, he felt it his duty to transfer not only his mining expertise but also his American notions of political liberty, good government, and free markets to British South Africa.[35] Yet his political misadventures revealed that market, state, and social relations greatly differed in South Africa's colonial environment. Exchanges of ideas and technologies were more often useful for the issues they raised than for their practical applications.

This gap between theory and practice often kept engineers from making their projects serviceable to their host cultures. Other obstacles stood in the way as well. First, historical contingencies—economic downturns or war—affected technical transfers in unpredictable ways. The Anglo-Boer War, for example, temporarily halted Californians' gold mining projects in South Africa. Second, environmental conditions often stymied engineers' attempts to tunnel underground into veins of gold or carry water over deserts. Floods, droughts, and other natural phenomena frequently impeded work; to their surprise, engineers found that they could not always control the vagaries of nature. Third, material development such as mines or irrigation systems theoretically helped pull a region into a global economy. Yet as the case studies of Australia, South Africa, Hawaii, and Palestine indicate, railroads and local markets had to be in place to enable regions to compete globally. Finally, the engineers' personalities and positions created further unevenness to their projects. Hammond painted an exalted portrait of the profession and its social and political leadership potential in his 1921 monograph, *The Engineer*. But more often than not, engineers were, as revolutionary thinker Thorstein Veblen argued in *The Engineers and the Price System* (1921), classic middlemen, caught between the demands of labor and capital, science and business.[36] Engineers' naive and sometimes utopian beliefs in their superior knowledge and technology led to frequent clashes with local bureaucrats, workers, and residents. In many instances, engineers allowed the profit motive to prevail over society's general welfare. Taken together, these factors ensured that engineers' projects often led to results far removed from the ideals they championed.

California engineers and their expertise circulated in many directions as the larger cadre of technocrats crisscrossed the globe. Californians initially looked to British India, the subject of chapter 1, as a global model of hydraulic development. British engineers in India constructed the world's largest canals, which were intended to alleviate famine and supplied British markets for cotton and grain, though Britain's export policies aggravated periodic famines in some regions. California engineers working overseas in the 1870s admired their British counterparts' seemingly absolute power over local peoples and landscapes but realized that this centralized authority could never be imported to California, as chapter 2 shows. In the United States, a strong civil society and laissez-faire economy hindered state intervention in irrigation development. California's engineers nonetheless modified the Indian model in ways that combined local, private irrigation development with state aid. They also designed other innovative resource policies, institutions, and agricultural settlement plans.

In the late nineteenth century, Californians shipped these models around the world. Their irrigation technology, land and water laws, and elite irrigation colonies found popular support in Victoria, Australia, discussed in chapter 3. Engineers' transfers might have succeeded in Victoria's similar political and cultural environment but for economic depression, lack of markets, and poor transportation systems. Similarly, Californians could not successfully export the state-aided settlement plans they had developed in Victoria back to California. The tools of progress adapted poorly in South Africa, examined in chapter 4. Building a modern economy meant creating social hierarchies involving elite foreigners—British mine owners and their California experts—and Boer and native black workers. Although Californians transferred their mining and irrigation tools, the colonial environment remained hostile to ideas about free markets and political liberty. Chapter 5 explores how California engineers and capitalists drew the Hawaiian Islands' sugar economy into world markets. Yet the growth of sugar plantations created new social hierarchies among Americans, imported Asian labor, and native Hawaiians; in some ways, the politics that underlay the idea of racial hierarchies helped drive the development of new agrarian structures. Attempts to rehabilitate Hawaiians on their land subsequently failed. Despite the varying social costs in the different locales, economic growth in India, California, Australia, South Africa, and Hawaii stood in marked contrast to Palestine, discussed in chapter 6. There, Jewish settlers under British rule found that their socialist ideals hindered the growth of an agricultural infrastructure that might have accelerated Palestine's further entry into world markets.

A key group of California engineers—some of whose accomplishments are well remembered in the historical records and others whose efforts are not—was instrumental in circulating technology and ideology to these different regions. Central to this itinerant community was William Hammond Hall (1846–1934). As California's first state engineer in the 1870s, he attempted to revise the state's antiquated water laws and design a comprehensive system for the state. Unsuccessful, he later brought his ideas to South Africa, where he joined his famed cousin, John Hays Hammond (1855–1936), in working with British mining magnates to develop reliable water supplies for the mines and for the growing city of Johannesburg. The Canadian-born George Chaffey (1848–1932) and his brother, William (1856–1926), built California-Australia connections in the 1880s when they tried to replicate the ideals embedded in their Southern California irrigation colonies in Australia. Finally, Elwood Mead (1858–1936), perhaps the most renowned engineer of this group, forged links among California, Australia, Hawaii, and Palestine. Best remembered for heading the U.S. Bureau of Reclamation between 1924 and 1936, Mead had previously drafted pathbreaking water laws in Wyoming and Colorado and chaired the State Rivers and Water Supply Commission in Victoria, Australia. In 1911, he returned to the University of California to lead the California Land Settlement Board. In the 1920s, he traveled to Hawaii and then Palestine, where he attempted to transfer his ideas about modern rural development and life. Recognizing geographic and other similarities between California and distant frontier societies, this cohort of engineers attempted to transfer the ideals of progress they had implemented in California to other parts of the world, with varying degrees of success.

What now appear to be the inevitable contradictions between the theory and the reality of spreading progress were simply not evident to most people at the time. Most nineteenth-century engineers in this global community did not consider the effects of their transfers of tools as we would today. They tried to transform arid regions into irrigated gardens, blast gold out of mountains, and build canals through deserts. They also believed that they could change society. Yet the spread of California's engineers and experiences and their reformulation in different regions showed that countries put foreign tools together according to their specific historical contexts. In turn, institutional arrangements such as land tenure, labor systems, and political and cultural values determined the applicability of exported technology. We now realize that technology cannot by itself alter cultures, social relations, governments, or economies—though it is coming ever closer to doing so. But California's engineers thought their tools would and ought to create such change.

The idea of progress in all its permutations gripped the nineteenth and early twentieth centuries in a way that is remarkable. This notion easily posed one of the most unchallenged, unexamined intellectual projects of modern times. Yet far from fulfilling a universal set of ideals, progress brought about terrible inconsistencies in some regions and unexpected consequences in others. Common themes nonetheless evolved in all of the regions that Californians visited. Colonial societies and nation-states alike deployed progress as a euphemism for environmental conquest, imperialism, ethnic cleansing, Anglo-Saxon and European superiority, and Social Darwinism. (Herbert Spencer, himself an engineer, argued that people and society followed the same basic scientific laws as the field of engineering.) Progress had deeper foundations in rhetoric than in reality and served certain people and goals at the expense of others.

Our understanding of these incongruities is not, however, as complete as we would like to believe. The theme of California engineers working in a global economy during the height of imperialism and technical innovation resounds as loudly today as it did in the late nineteenth century. Racial and ethnic hierarchies, for example, have characterized all of human history, from the ancient Greeks, Persians, and Egyptians to imperial Britain, the American South, and the modern Israeli state. Equality, a relatively new concept, has yet to withstand the test of time. Human-induced environmental change has likewise been constant throughout human history. Although nineteenth-century ideas of conquest and domination over nature yielded to "sustainable development" in the late twentieth century, environmental change still advances at a frightening pace. Finally, we continue to imbue technology with cultural values and to allow those values to guide our actions.

If our checks on these processes differ from those of a century ago, the concepts have hardly changed. Twenty-first-century globalization—particularly as enabled by the Internet and the personal computer and by the gradual shifts that individuals and companies around the world must make to compete in a global market where geographical and historical divisions are becoming less important—just may be a newer, faster version of the older circulation of money, tools, and ideas. This process has yet to "flatten" the world, as Thomas Friedman and others argue. But modernization, as evidenced by many regions in present-day Asia, Africa, and the Middle East, still takes more than strongmen, technology, and money. McDonalds, Cadillacs, and Levis—as well as telecommunications companies, innovative startups, and desktop labor—cross borders, but free markets, political parties, and the rule of law do not always follow. We are so tuned into global capitalism that we cannot fathom just how revolutionary these forces remain in many cultures. This historical perspective on the roots of globalization allows us better to understand why Californians thought and acted as they did abroad.[37]

Despite its utopian rhetoric and displays, the 1915 World's Fair broke sharply with the past. World War I momentarily dislocated humanity's confidence in the ideals that had self-consciously guided the nineteenth century. It marked a time when no ship safely crossed the Atlantic, nations fought nations, and engineers designed U-boats and barracks. But the dream of progress, if fractured by war then and many times since, recovered its self-confidence. In the twenty-first century, it remains as strong a concept as it has ever been. Hammond concluded his 1935 autobiography with words that echo as loudly today as they did in the nineteenth century: "Our machine civilization has been wrought by the engineer, who contrives its apparatus, utilizes and harnesses the physical and chemical forces of nature, and exploits the resources of the earth. All that he designs and invents and exploits redounds in the end to the public benefit."[38]

### CHAPTER 1

# Lessons of Valuable Experience

WHAT CALIFORNIA LEARNED FROM INDIA

The American government chartered or subsidized many of the nation's early enterprises, including interstate railroads. Yet it showed little support for irrigation development in California, even when agriculture began to surpass mining as the state's predominant industry in the 1870s. Individuals and private companies built ditches and canals to carry water to their dry fields. Such unregulated private development, while dynamic for the mining industry, produced chaotic results for irrigation. California's water laws allowed irrigators to monopolize water to the detriment of other users. The shortage of capital and labor as well as battles among miners, farmers, and ranchers also hindered irrigators' projects.

With agriculture booming, irrigators started to cry out for government aid. To establish the scientific and technical feasibility of irrigation, public officials examined private irrigation works in California's Great (Central) Valley and discussed ways to aid farmers and irrigators. State leaders then conducted comparative inquiries into the activities of other countries, particularly India, and returned home to think about ways to apply what they had learned abroad.

### The Private Canal Company and the Federal Survey in California

Between the 1850s and 1870s, the mining industry spurred the state's rapid population growth and created a dynamic, barely regulated economy that in turn further fueled the expansion of mining. Weak government oversight, however, negatively affected the growth of irrigation. The example of the San Joaquin and King's River Canal Company illustrates the attempts to involve the government in private irrigation projects in the early 1870s. Although the canal company did not acquire government subsidies, it sparked policy makers' interest in irrigation in California. The first federal survey of the Great Valley, led by Barton S. Alexander and California engineer and scientist George Davidson, initiated discussion about the role of public institutions in resource development but failed to reconcile the state's haphazard water laws and irrigation systems.

An open society and economy characterized California from the state's inception. The gold rush gave entrepreneurs the capital and engineers the technical experience to assemble a modern infrastructure within three decades. The discovery of silver in Nevada's Comstock Lode in 1859 transformed San Francisco from a frontier outpost into a growing international port. Merchants sent gold, dredges, drills, and men to distant places, while the city welcomed vessels with Hawaiian sugar and goods from China and Japan. William C. Ralston's Bank of California, the San Jose–San Francisco Railroad, and municipal fire departments, wharf commissions, and newspaper agencies settled in. A rail and water network connected San Francisco to the rest of the country a decade later. But the gold rush did not create a political framework for the development of California's natural resources. Gold attracted settlers but also unlocked the state for unbridled exploitation. Anglo-Saxon settlers rapidly extinguished Indian claims and Mexican land titles, chased competitors off the newly claimed land, and blasted away at California's mountains.

Placer (surface) mining, which required few major water diversions and kept waterways relatively intact, dominated the Sierra Nevada's gold industry for the first few years after 1848. Hydraulic mining, a more capital-intensive and destructive method of unearthing ancient veins of gold, replaced placer mining by the late 1850s. Hydraulic mining required a system of canals and water storage for continuous operation. Miners used pressurized hoses to blast hillsides and then transported the rock debris by flume for processing downstream. Between the 1850s and 1870s, engineers constructed large rock-fill dams, such as the one-hundred-foot-high English Dam and ninety-six-foot-high Bowman Dam in the northern Sierra, to supply water to the mines. By 1867, more than five hundred miles of canals and ditches channeled water from Northern California's Sacramento, Bear, Feather, and Yuba Rivers to miners.[1]

Despite the rapid proliferation of mines, the state neither supervised nor regulated the industry. "The policy of the state should be to legislate as little as possible on the mines," noted the California Senate Committee on Mines in 1856, echoing the general sentiment.[2] Corporations did not even have to secure ordinary business licenses to tear down mountainsides. Instead, hydraulic mining companies shaped government policy to their own interests. The state, for example, exempted white miners from taxes and adopted the Foreign Miners Tax in an attempt to eliminate Asian and Mexican competitors.[3]

California's previous groups of inhabitants had managed land and water in a completely different way. Native American hunter-gatherers had seasonally roved California's forests and streams, to which these inhabitants exerted usufructuary rather than exchange-value rights. In the eighteenth century, Spanish missionaries tried but failed to use Indian labor to cultivate small patches of irrigated agriculture to support the Spanish forts, missions, and pueblos. Until the 1820s, the Plan of Pitic, which established common law for waters and privileged community over individual use, governed all water rights. The Mexican rancho period between the 1820s and 1840s similarly adopted the pueblo, or community rights, doctrine.

Mining changed these patterns of community control. William Hammond Hall, who served as California's first state engineer in the 1870s and 1880s, pointed out that gold rush California presented a unique situation. Thousands of settlers quickly flooded the state, and few were aware of any water laws. Both the stakes and the possibilities for immediate wealth were tremendous.[4]

Yet this laissez-faire attitude toward mining confused water claims. The doctrine of prior appropriation — "first in time, first in right" — emerged out of the gold fields, where miners diverted water far from rivers' banks. Water became private property removed from property in land. The principle of prior appropriation contrasted with an earlier riparian law that the state had implicitly adopted in its 1850 constitution. This riparian law, an ancient Justinian public trust doctrine prevalent throughout the rest of the country, allocated water to people living on the banks of a river.

By the late 1850s, prior appropriation predominated de jure, although it clashed continually with the older riparian principle. Prior appropriators and riparian owners alike monopolized water, either alongside other people's land or to the detriment of downstream farmers.[5] Ambiguity at best prevailed in water matters. Irrigators, for example, filed seventy-six claims on the Kern River under different laws and diverted water from thirty-two canals where five would have sufficed. Even if irrigators had streamlined their water withdrawals, they still would have found something to gripe about: the volume of water they claimed exceeded the river's flow twenty times over. Very rarely did judges pacify sparring parties or amend laws.[6]

California's water laws provided the greatest obstacle for irrigators, but other problems existed as well. The state's low population density prevented water projects from being profitable. As labor shortages kept the cost of construction high, farmers built wooden structures with an eye toward economy. "Irrigation has grown up in a haphazard way," engineer Frederick Haynes Newell of the U.S. Geological Survey commented, "with little or no preliminary examination [and] having more regard for quick returns than for future permanence of works of water supply."[7] Upon visiting California in 1885, Alfred Deakin, the minister of water supply in Victoria, Australia, similarly remarked that "extreme simplicity and economy" dominated California's irrigation works "whenever it can be shown to be profitable." In the Mussel Slough area of the San Joaquin Valley, Deakin observed that farmers dug canals with teams of horses pulling Fresno scrapers and primitive cast-iron scoops and did not line the canals.[8] For good reason, he called irrigation in California "child's play."[9] To his horror, he observed that the government seemed indifferent to water planning and conceded that the "utmost political aspiration of the American irrigator is that the Government will leave him severely alone."[10]

Deakin's observation pointed to the truth. While British engineers excavated channels and laid stones for the world's largest canal system in India, California politicians had yet to fund irrigation schemes. For the first two decades of its existence, the state government, at least in the area of agriculture, did little more than dole out land and information to farmers. The federal government disposed of land in a series of acts that sought to carve yeoman empires out of arid lands. But many of these acts, including the Homestead Act (1862) and Desert Lands Act (1877), rewarded speculators. The Swamp Land Act (1850) allowed speculators to acquire large tracts of land in the San Joaquin Valley. Such legislation seemed to suit the needs of a relatively unpopulated state. Yet as settlers poured into California after the gold rush and agriculture began to replace mining, it became clear that private enterprise could not alone develop California's ample resources.

Several factors conditioned California's great agricultural leap between 1850 and 1870. Gold production declined after 1854. Many settlers left the Sierra Nevada's gold mines for farming. Agriculture expanded, first in response to mining and then as an industry in its own right. The 1846 repeal of the Corn Laws increased Britain's reliance on imported grain. New markets in Liverpool and food shortages all over Europe, combined with the huge population increase in America's eastern cities after the 1840s, created a high demand for livestock and grain products. Modern farming machinery and the growth of transportation systems also spurred agricultural growth.

During the 1860s, farmers began to grow wheat and barley in the dry San Joaquin Valley, the Great Valley's southern end, for large-scale export to New York and Great Britain. These grains adapted so well to the valley's fertile soil and dry, treeless clime that surpluses soon developed. Merchants floated grain down the Sacramento and San Joaquin Rivers to Red Bluff and Stockton, which had access to riverboats running to San Francisco and the northern mining camps. Grain production skyrocketed. Between 1866 and 1869, the number of acres planted in wheat increased from 89,563 to 380,547. The Southern Pacific Railroad began construction of a line through the San Joaquin Valley in 1870.[11] The railway as well as the demands of the state's population, which mushroomed from 15,000 in 1848 to 360,000 in 1860, stimulated even more production. By 1870, the counties in the Sacramento and San Joaquin Valleys numbered among the world's leading grain-producing districts.[12]

Lack of rainfall limited agricultural expansion, however. The Great Valley runs lengthwise through the state, perpendicular to the western Sierra Nevada mountains and the eastern Coastal Range. It receives the melting snows of the Sierras but averages only fifteen inches of rain annually, not enough to sustain year-round crops. Prolonged droughts between 1863 and 1864 and between 1868 and 1872 devastated the San Joaquin Valley's cattle and wheat industries.[13]

Many Californians believed that people could change these natural conditions with experimentation, technology, and money. In 1852, capitalist John Bensley arrived in San Francisco and founded the California Steam Navigation Company, a joint stock venture that controlled the bulk of the traffic on the Sacramento and San Joaquin Rivers. Five years later, he organized the San Francisco Water Works Company. What would happen, he wondered, if water could be brought to the San Joaquin Valley every day of the year and drained away during wet periods? Bensley realized that irrigation would grant the Great Valley's large landowners a monopoly on wheat.

In 1866, Bensley filed claims on a large part of the San Joaquin River and organized the San Joaquin and King's River Canal Company. Hiring one thousand Chinese laborers to build a canal along the river at Firebaugh's Ferry, he then leased farmland along the canal route and put it under cultivation to show the feasibility of irrigation in the valley.[14]

The high costs of labor and of transporting equipment seventy-five miles over the Coastal Range drained Bensley of capital. He thus turned for help to his old friend Ralston, the mastermind behind Nevada's Comstock Lode, president of the Bank of California, and owner of San Francisco's Spring Valley Water Company. Ralston poured millions of his own money and an unauthorized million of his bank's money into Bensley's canal company.

Appointed president of the company, Ralston then solicited backing from a set of San Francisco's movers and shakers: William S. Chapman, the single-largest landholder in California; Chapman's partner, "Grain King" Isaac Friedlander, who ran a large wheat export trade with England; Nicholas Luning and A. J. Pope, directors of the Bank of California; Lloyd Tevis, president of Wells Fargo; and James Ali Ben Haggin, a large land-holder in the valley. The company also had two silent partners, Henry Miller and Charles Lux, stockmen who owned the Santa Rita Ranch and thousands of acres on the west side of the San Joaquin River below Fresno Slough. They owned so much land and water in the San Joaquin Valley—450,000 acres of land and 120 miles of riparian water rights along the San Joaquin River—that a canal by necessity had to pass through their property. Miller and Lux promised to provide rights-of-way in exchange for company stock. They also pledged to pay twenty thousand dollars toward the cost of construction over three years and to purchase adequate water rights to irrigate more than 100,000 acres of pasture.[15]

Ralston had ambitious plans. He envisioned a 230-mile aqueduct running from Buena Vista Lake in the Coastal Range to Tulare Lake and then to Antioch, at the northern end of San Francisco Bay. Eight lined canals would link the main canal with the San Joaquin River, transporting water from the northern end of the basin, where most of the rain fell but only one-third of land was arable, to the southern, drier end. The company's directors predicted that the main canal could irrigate one hundred thousand acres of grain and grasses. Aqueducts would transport boats carrying farm machinery, grain, and building supplies and provide water for irrigation across the state.[16]

Ralston solicited the engineering expertise of Robert Maitland Brereton, an internationally acclaimed British engineer who, Ralston believed, could design such a system. Brereton had amassed considerable experience in India, where he had designed the Indian Peninsula Railway from Bombay to Calcutta and Madras in the 1850s. He had then studied India's extensive British irrigation systems. Ralston lured Brereton to California in 1871, despite competing offers from the Japanese government. As the company's chief engineer, its business manager, and a forty-thousand-dollar investor, Brereton became central to Ralston's scheme.[17]

Brereton saw many parallels between California and India, particularly their hot winds, arid climate, and dry, dusty soil—and many differences. British engineers relied on corvee labor to construct government-owned irrigation works such as canals, which sought to alleviate famine for India's teeming masses and opened up regions to new markets. California's Great Valley, by contrast, experienced a shortage of both labor and settlers.

Brereton nonetheless encouraged Ralston to model the company's water distribution network after the British system, to no avail. A few of the company's directors wanted to build a main canal and leave the construction and cost of distribution ditches to farmers. Brereton warned, however, that "if these distribution ditches be left to 100 different parties to carry out *without the control of the company*, each anxious only for his own interests . . . the full advantages of irrigation will be considerably reduced."[18] He also cautioned that landowners Miller and Lux might one day wield their riparian rights against downstream water users. By contrast, the British government owned all the rights to northern India's large rivers.[19]

The San Joaquin and King's River Canal Company completed thirty-eight miles of aqueduct by the end of 1871, just as an economic depression swept through California.[20] Ralston sent Brereton to England to find more investors the following spring. When he returned without the desired investments, the company petitioned for federal support. In 1872, Brereton traveled to Washington, D.C., and lobbied President Ulysses S. Grant, Speaker of the House James C. Blaine, General William Tecumseh Sherman, and Ralston's old friend, Nevada senator William Morris Stewart, for two separate but related pieces of legislation.[21] The first was a special interest item aimed at obtaining for the company land grants and rights-of-way across public and private domain in the Great Valley, much as the railroads had received. The second bill sought an appropriation for the appointment of a board of commissioners to report on a system of irrigation for the San Joaquin and Tulare Valleys.[22]

Brereton's first bill was introduced in the U.S. Congress on January 17, 1873. The bill's provisions included rights-of-way through eight counties; two sections of public land per mile of canal and 100 acres per mile to pay for the construction of reservoirs; and a monopoly over the waters of Buena Vista and Tulare Lakes and parts of the Kern and San Joaquin Rivers—a total of 256,000 acres.[23] Brereton argued that the public stood to benefit if the government condemned all privately held lands surrounding the proposed canal, but detractors argued that land grants led only to monopoly.[24]

A month after the bill to cede land to the San Joaquin and King's River Canal Company failed, Senator Stewart introduced a bill for a federal irrigation survey in California's Great Valley. In March 1873—five years before John Wesley Powell published his famous *Report on the Lands of the Arid Region of the United States* and a quarter century before the birth of the U.S. Bureau of Reclamation—Congress committed its first funds toward an investigation into the valley's irrigation and reclamation possibilities. Intrigued by the canal company's project, Congress desired a technical and scientific report on the best system of irrigation for the Sacramento and San Joaquin Valleys.[25]

The federal irrigation survey, which became known as the Alexander Commission, took shape as an outgrowth of the San Joaquin and King's River Canal and Irrigation Company. The company no doubt anticipated that the federal commission would recommend a land grant and seal the partnership between California's grain kings and the government; Stewart added the Sacramento Valley to the scientific examination of the San Joaquin Valley only after the recommendation of another senator. The legislation also created an irrigation commission that included Lieutenant B. S. Alexander and Major George Mendell of the Army Corps of Engineers; Davidson, of the U.S. Coast Geodetic Survey; and Brereton.[26]

All of these men were trained engineers who in many ways bridged the gap between the scientist-engineer of the nineteenth century and the "objective" managerial expert of the twentieth.[27] Alexander, a native of Kentucky, had graduated from West Point in 1842 and had headed the Union Army's military engineering under President Abraham Lincoln. After the Civil War, Alexander traveled to the West Coast, where he was appointed senior military engineer of the Army Cops of Engineers. Mendell, also a West Point graduate, joined the Topographical Engineers in 1852. For the next eight years, he surveyed a railroad route from San Francisco to Yuma, Arizona; led campaigns against Indians in the Oregon and Washington Territories; and supervised improvements to San Francisco Bay and other Pacific Coast harbors.[28]

Davidson, who wrote the Alexander Commission's final report, was a renowned scientist, engineer, and geographer. A Scot by birth, Davidson came to the United States as a boy. In 1850, after graduating from the University of Pennsylvania, he began a long career with the U.S. Coast and Geodetic Survey. He believed that California's gold rush would increase waterborne commerce. Trade, in turn, would necessitate the accurate mapping of California. For the next three decades, Davidson headed the survey's Pacific Coast work from its San Francisco office. He conducted pathbreaking telegraphic longitudinal work and delved into various aspects of hydraulic engineering, harbor and river improvements, and urban and rural water supplies. He also served for many years as president of the California Academy of Sciences and professor of astronomy and geography at the University of California in Berkeley.[29]

The link between the federal survey and the San Joaquin and King's River Canal and Irrigation Company did not escape the notice of small farmers in the San Joaquin Valley. Like Congress, they questioned the durability of the stretch of canal completed by 1871. Others feared collusion between public agencies and private capitalists. Settlers in the valley, the *Pacific Rural Press* claimed, unanimously opined that "the water should belong to the State or Government, and if they must pay for it, they do not want to enrich monopolies by doing so."[30] They thus echoed the sentiments of Henry George and other contemporary critics of land and water monopolies.

Small farmers realized that their livelihoods depended on their ability to mobilize scarce state funds. Like miners and merchants, farmers created associations to look after their special interests, including the California State Agricultural Society, the Agricultural Society of Northern California, the California Farmers' Union (and its successor, the Grange), and others. These associations voiced farmers' beliefs that the state government should do more to disseminate information on agriculture, enact fencing laws to keep cattle out of farmers' fields, oversee shipping rates, promote railroad competition, and help farmers deal with mining groups and cattle raisers.

Many of these lobbying efforts failed to bear fruit. The legislature gave direct grants to farmers and exempted many from taxes or liens to safeguard their property, as it had done for miners. But the state lacked the funds to provide adequate administration.[31] Competing interests also fractured farmers' lobbying efforts. Unity was scarce between Northern and Southern California agricultural interests, irrigating and nonirrigating farmers, richer and poorer farmers, and proponents of different types of farming.[32]

Davidson sympathized with the plight of the small farmers and believed that irrigation—science applied to industry—would solve some of their problems. Viewing this civilizing work with the optimism of many nineteenth-century experts, he predicted that irrigation would transform California's Great Valley into the world's breadbasket within two decades. Twelve million acres could be cultivated—8.5 million acres in the Great Valley, including swampland, plus 3.5 million that included the foothills. The federal commissioners projected annual returns of fifty to eighty bushels of wheat and five crops of alfalfa per acre.[33]

The Alexander Commission met in May 1873 and conducted most of its survey work that summer. Yet the commission fell short of its goals. The government allocated a meager five thousand dollars toward the project, too little for the commission even to draw maps of potentially irrigable lands. Davidson admitted that the project would take more time than could be feasibly handled by men who already held full-time jobs. The commissioners did, however, measure the flow of some rivers and streams, designate the locations of potential irrigation canals, and define the natural boundaries of a few watersheds.[34]

The commissioners also outlined a statewide irrigation system that depended on gravity and the state's topography. Both ends of the Great Valley, Red Bluff to the north and Bakersfield 450 miles to the south, lie several hundred feet above sea level. The commissioners foresaw that the east side of the valley, which receives the snow from the Sierra Nevada mountains, could easily irrigate the drier west side. They suggested building one major aqueduct that would connect the Sacramento and San Joaquin Rivers by a series of canals as well as several small dams.[35]

The Alexander Commission's report disappointed its expectant beneficiaries. The San Joaquin and King's River Canal Company had no doubt anticipated that the commission would recommend that Congress issue a land grant and would seal the partnership between California's grain kings and the government. To the canal company's chagrin, the commissioners understood the scientific risks of large irrigation projects and failed to endorse the scheme.[36]

The commissioners not only questioned the feasibility of the project but also, and more important, feared collusion between the government and the canal company. For Davidson, Ralston's canal company symbolized the state's faulty land and water laws. The valley's big landowners—Chapman, Miller, and Friedlander—regularly used their political influence to secure low assessments on their land, thus increasing the tax burden on the rest of the farming population.[37] The commission believed that the company, if given land grants and subsidies, would follow the monopolistic pattern of the Southern Pacific Railroad, which, along with Miller and Lux, had already snatched up much of the Great Valley's valuable land and water rights.

Davidson nonetheless understood that individuals could not undertake large-scale irrigation projects and believed that the government had a duty to teach the value of irrigation, design a comprehensive irrigation system, and enforce proper laws.[38] Yet Davidson's colleagues felt that a large grant to the San Joaquin and King's River Canal Company was not appropriate until either the federal or state government was able to supervise construction of reclamation works, regulate water rights, and prevent waste by determining the amount needed to irrigate a certain area.

Instead of endorsing Ralston's canal scheme, the Alexander Commission's recommendations affirmed members' vision of a nation of small farmers aided by the state. The commissioners first suggested a partnership between private and public groups whereby the state and nation would

jointly undertake topographical and hydrographic surveys, locate potential irrigation areas, and design a network of canals and aqueducts. Private
companies, supervised by state officials, would construct the systems, and
state-chartered irrigation associations would distribute the water. After
aqueducts carried water to the Great Valley, the increased land value would
offset the initial high cost of irrigation.[39]

The Alexander Commission also recognized the need for reform of California's water laws. Davidson firmly believed in the riparian principle and
thought that legislators could prevent land and water monopoly only by
attaching water rights to the land. The "nonsuccess of private organizations
for the purpose of carrying out great irrigation schemes," Davidson wrote,
"where the owner of the land and the owner of the water are different
parties, or where there are conflicting claims to water, may be enunciated
as almost a law of commercial investment."[40] "When the water is thus given
to the land it should be in perpetuity," he argued. He saw an indissoluble
bond between land and its adjacent water, "the fee simple of one to the fee
simple of both."[41] In other words, owners could possess both resources outright. Solving these issues, however, required a complete overhaul of California's existing water policies.

The government's unwillingness to lend direct support to Ralston's San
Joaquin and King's River Canal Company had calamitous results for the
company, which had run out of money by the end of 1873. In December,
Ralston tried to sell the company's holdings to the state, which refused the
offer. To make matters worse, Chapman, Friedlander, and Ralston's Bank of
California went bankrupt in 1875. The following August, bank directors de-
manded Ralston's resignation based on suspicions regarding his solvency.
That same day, Ralston ventured to the public beach for his daily swim in
San Francisco Bay and drowned.[42]

Brereton mourned that this turn of events had not only "killed that
great and first enterprise in irrigation in North America" but also ruined
his career, since he had reneged on his appointments under the government
of India. Unfavorable conditions forced Brereton to sell his shares of canal
stock to Miller and Lux for only one thousand dollars. The cattle barons also
acquired the company's best land and the unfinished canal, which they subsequently extended twenty-seven miles into cereal and alfalfa fields.[43]

The company's project failed for many reasons. The Great Valley's low
population density offered little support for the project. High idealism
masked poor engineering skills and the absence of the evidence needed to
convince the larger population of the benefits of irrigation. Similarly, California's water laws—in particular, the doctrine of prior appropriation—

hindered the project. As historian Donald Pisani, whose history of this period of irrigation remains a definitive account, writes, "One of the nineteenth century's boldest dreams died with Ralston," and his plan "to reclaim the entire Central Valley through private enterprise would never be revived."[44]

The Alexander Commission's vision met a similar fate. The commissioners greatly underestimated the time and planning required to reclaim the Great Valley. Their report also elicited hostility from politically influential San Joaquin Valley cattlemen, who owned large tracts of land and used streams exclusively for grazing purposes.[45] Far from drawing a comprehensive blueprint for a statewide irrigation system, the commission's report captured the problems caused by unregulated irrigation development. Yet despite all of the unfavorable indicators of California's irrigation future, Davidson remained optimistic. "Although the life blood of the country is running through our Valleys to the ocean," he wrote, "it may be made to water every available acre. Other countries have succeeded—we shall compel success. They have had a thousand difficulties—we must profit by their troubles and mistakes, or inflict upon ourselves still greater complications."[46] Davidson thus turned to India.

### We Must Profit by Their Troubles: George Davidson's Visit to India

As the birthplace of modern hydraulic engineering, India served as an international school for civil and hydraulic engineers. Deakin, who visited India in 1890, five years after his visit to the American West, explained that "India has run the whole gamut of irrigation works, and is therefore incomparably superior . . . to any country in the world."[47] In the nineteenth century, British engineers began to transform ancient inundation channels into modern canal systems.[48] Davidson visited India in 1875, more than a year after he had served on the Alexander Commission. He admired India's works and viewed the strong British state as a possible solution to the chaos of private enterprise in California. Davidson thus looked to India as a model for irrigation. However, he overlooked the difficulty of importing India's hydraulic regime, with all its social and political inequalities, into a democratic society.[49]

Popular, scientific, and government articles on Indian irrigation started to appear in the American press at midcentury.[50] *Scientific American*, for example, hailed India's ability "to correct the unkindnesses of Nature" with waterworks "remarkable for their bold conception" and design.[51] The U.S. government also solicited information about India; in 1849 and for nearly every decade thereafter in the nineteenth and early twentieth centuries, the U.S. Congress sent consuls abroad to inquire about other countries' irrigation regimes. Americans reported on Mexico, China, Siam, Australia, South Africa, Syria, Palestine, and the Sandwich Islands, but Americans had unrivaled admiration for India's efficient and successful irrigation techniques.[52]

In 1874, Davidson left San Francisco for a trip around the world. He was one of the first of his American colleagues to travel to distant places to meet engineers, study water laws and technologies, and bring home new ideas.[53] He inspected irrigation works in China, Egypt, India, and Italy, but these visits represented a mere side trip to his more pressing interest of observing the transit of Venus across the face of the sun in Japan. The trip was funded by the U.S. Navy and Treasury Departments and may have been arranged at the behest of the San Joaquin and King's River Canal and Irrigation Company.[54]

Davidson traveled from Bombay to Suez in fifty days and compared the systems of irrigation, reclamation, and engineering in relation to roads, railroads, forests, sewers, and water supply. He noted the absolute necessity of artificial irrigation in Egypt. The Nile Valley afforded ample transportation from Cairo to Suez but lacked regular rainfall. Canals carried water from the Nile River to the valley's fertile soils. Yet irrigation engineering was still in its infancy, and Davidson wished only to avoid some of Egypt's mistakes. Japan's ample rainfall and terraced waters had no corollary to the drier parts of the United States. Chinese irrigators relied on old methods and customs. France, Spain, and Italy enjoyed relatively advanced water laws but bolstered Davidson's belief that private companies could not successfully undertake large-scale irrigation works.

Davidson covered thirty-eight hundred miles over India's rough terrain in eight weeks. He examined twelve large public waterworks constructed, managed, and owned by the British imperial state. The country, he reported, stood "pre-eminent in her gigantic undertakings" and "unparalleled" engineering masterpieces. Nations built works on grand scales, but India eclipsed even Egypt in its scope of vision. Davidson admired dams grander than the pyramids and massive stone bridges decorated by carved lions, the symbols of the British Empire.[55]

The British boasted the most modern hydraulic regime in the world. Bombay's Bhatgur Reservoir alone held 119,600 acre-feet of water and rose 130 feet above its foundation.[56] The Báree Dóab Canal, with its headworks at the foot of the Himalayas in the Punjab, the cradle of the Sikh nation, crossed two small mountain ranges, struck into "wild and arid wastes," and regenerated "ruined cities and villages" for its 247 miles.[57] Both systems surpassed anything Davidson had ever seen in California. The extensive Ganges and Jumna systems dazzled him even more.

"If we wish to profit by the necessities and dearly bought experience of other countries," Davidson wrote, "no field is so suggestive to the American mind as that of India."[58] The Punjab and Indus Valley had close geographical counterparts in California's Great Valley. The Indus, Jumna, and Ganges Rivers resembled the Sacramento, San Joaquin, and Feather Rivers.[59] California and India were also two of the few semiarid places in the world that boasted large swaths of fertile soil. The "qualities of the English governing race in India," Davidson also observed, "are almost identical with those of our own people." This cultural heritage further suggested the supremacy of the Anglo-European race in bringing progress to the barren parts of the world.[60]

Davidson failed to see that the imperial-colonial relationship undermined the material progress he observed. He met with British engineers and saw their monumental work but did not record his thoughts about local villagers' roles in these undertakings. He mistook state power for progress and erroneously believed that the British constructed grandiose irrigation works to save loyal subjects from famine and drought.[61]

Davidson arrived in India shortly after passage of the controversial Northern India Canal and Drainage Act of 1873, which granted the British government the supreme right to claim and use water in northern India. The Bombay presidency for Sind, the southern half of the Indus Valley, adopted similar legislation six years later. The 1873 act reconciled private and public rights, giving the government full ownership and control over all water and the power to regulate irrigation, navigation, and drainage. The measure also authorized imperial officers to enter private lands, levy water rates, and construct works. Water rights inhered in the land.[62] The act elicited Davidson's admiration for India, seeming to clear out a host of competing and costly interests. The legislation also convinced him that only a strong state could invest in and profit from irrigation and organize an expert bureaucracy to manage irrigation ventures.[63]

The connection Davidson made between centralized state power and irrigation had its roots in ancient empires. After visiting China in 1938, Karl Wittfogel, the architect of the hydraulic society thesis, wrote of the link between large-scale irrigation and political power. The despotic control of water projects in the river valleys of Egypt, Mesopotamia, India, and China and among the Pueblos and Zuni of the Americas suggested that large-scale irrigation and state power evolved together. Managerial bureaucrats built immense dams and aqueducts to support their empires in arid conditions, thereby becoming the ruling class. In most cases, this elite group of technocrats repressed constitutional and societal checks by civil society, usually a body of peasant corvee labor. Wittfogel argued that hydraulic societies repressed social mobility and thus the early expressions of capitalism spreading from Western Europe in the fourteenth century. A decentralized agricultural system under feudalism, together with abundant rainfall, had allowed capital to accumulate in Western Europe. These conditions produced the rise of a modern, property-based industrial society. But in hydraulic societies, the state usually restricted the development of private property. Although these societies boasted sophisticated irrigation works, they eventually collapsed when they depleted their resources.[64]

Irrigation in India, like Wittfogel's hydraulic societies, related closely to the structuring of political power, an argument Donald Worster makes in *Rivers of Empire*. Many British imperialists saw the construction and control of railroads, telegraphs, irrigation, and other infrastructure projects as a measure of economic power and a symbol of themselves as a ruling community in India.[65] Although the British East India Company and other private ventures had encouraged the cultivation of vast areas of land by improving on old inundation systems, the general agricultural conditions remained much as they had been for centuries.[66] The British realized that, short of population decline, only large-scale irrigation would alleviate recurring famine. Planned irrigation, facilitated by lands opened up by the railroad, would also open up India to world markets, settle the peasantry in the Crown's northern "waste" tracts, and help control a large, politically divisive region.[67]

Embedded within a large bureaucracy, British engineers started to build dams and hundred-mile-long canals across India in the 1850s, after repeated failings by private companies. The British effort started south of the Himalayas from the Indus River Delta in the west and curved east to the mouth of the Ganges River. Within decades, previously arid regions produced crops to feed the world's densest populations. In 1875, Davidson predicted that by 1900, "the population will be doubled, famines will be mitigated, the revenues of the government more than doubled, and the rule and authority of the English more firmly established."[68] Indeed, by the early part of the new century, government-owned canals stretched across more than thirty-nine thousand miles of desert. They carried water to an estimated 30 million of the country's 197 million total cultivated acres. Irrigation, Davidson observed, changed the "whole face of Northern India."[69]

The Ganges Canal, which served the famine-prone Upper Doab districts, impressed Davidson the most. The system boasted 70,000 masonry wells, a dam 4,300 feet in length, 568 miles of main canal, 3,293 miles of distributaries, and 895 miles of escapes and drainage cuts.[70] Each tributary eclipsed the size of America's major canals. The Ganges Canal yielded returns ranging from 9 to 12 percent on a capital outlay of four million pounds.[71] Other feats were equally impressive: the Agra Canal on the bank of the Jumna River, the Soane Canals near the East India Railway, the Sirhind Canal at Boopúr, and the Bari Doab Canal in the Punjab.[72]

The British performed their most impressive feat in the Punjab, a plain that stretches along the northern end of the great Ganges Valley. Bordering Afghanistan to the east and the Himalayas to the west, the province covers more than 80,700 square miles, an area equal to England, Wales, and Ireland together. The Punjab's rivers slope south from the Himalayas and unite in the Indus, the largest river, before leaving the area. In the dry season, the scrub-covered Punjab receives a mere two to ten inches of rain, not enough to sustain crops.[73] In the 1860s, the Irrigation Department, a branch of the Public Works Department, began to construct immense perennial canals and relocate peasants from good desert families.[74] Each peasant received one to two acres of government-owned land. By the 1890s, the Triple Canal colonies alone boasted four million plots of land for farmers. On his 1890 visit, Deakin observed that the Lower Chenab Canal Colony resembled the "colonies which private enterprise has scattered over Southern California and Colorado."[75] British engineers designed more than four hundred miles of main canals and twelve hundred miles of distributary canals to spread water over two million acres of crops.[76] Scientists tested the alluvial soils; engineers built canals and high masonry dams. Officials also divided the Punjab into units following the contours of natural waterways, laid out towns according to watersheds, and arranged water rights and distribution systems before colonists arrived.[77] Within a decade, irrigated fields of corn, indigo, wheat, cotton, and rice as well as roads, railways, towns, and markets sprang up, showing that in some regions, government intervention paid off.[78]

The Northern India Canal and Drainage Act created the basic structure for water administration in the region. The scope of public water control and the powers assigned to engineers astonished Davidson. The Irrigation Department of the Public Works Department monitored all construction, waters, and irrigated lands and had a vast hierarchical system of engineers.[79] But like most visitors, Davidson did not put the law's operations into the local political context. The act claimed complete state control over irrigation water, yet its provisions contradicted some customary laws relating to water that hindered British land settlement schemes in the Punjab and Indus Valleys. Davidson also failed to recognize that the act applied only to unappropriated waters and their adjacent lands. Thus, the power that the British government possessed over land and water was less than he assumed; state ownership of irrigation works did not mean complete control of all water and its adjacent land.[80] These limitations suggested a more serious flaw in India's hydraulic regime: the lack of partnership between water users and the state. The government acknowledged no local titles for water users and seemed to value local knowledge very little, though much local control over agriculture and tensions between a highly localized Indian water system and the more centralized British one persisted.[81]

Davidson observed other failures. British engineers constructed modern irrigation works with primitive machinery and compulsory labor. Davidson watched Indian men, women, and children fill, carry, empty, and return large baskets of stones in a way that would "drive an energetic engineer to despair." Although he advocated small farms, he predicted that Californians could increase the amount of work done sixfold just by using steam-powered equipment. He could not deny, however, that despite certain engineering mistakes, India's substantial masonry "would last through the next five centuries."[82]

The overall lack of scientific, ecological planning posed more serious problems. Davidson criticized the British for undertaking immense projects with "insufficient data" and for breaching "natural laws."[83] Overcultivated soil declined in fertility after just a few years. Abandoned wells saturated the soil and ruined crops. The first agricultural chemist to report on India's environmental conditions estimated that salinity had significantly decreased the value of almost five thousand square miles of agricultural land. Deforestation only compounded these problems.[84]

Yet Davidson worried little about the biggest blunder, the inequality generated by the imperial-colonial relationship and suppression of local self-government. Seeing a direct connection between the success of India's irrigation works and California's future, he believed that settlers could profit from the Indian example. Davidson's trip to India thoroughly convinced him that the "Indian system, with modifications, [could] doubtless be successfully applied" to California.[85]

## Assistance in Some Practical, Tangible Shape

Davidson's trip to India greatly influenced his ideas about the state's role in irrigation development. After returning to California, he published a report of his travels, *Irrigation and Reclamation of Land for Agricultural Purposes, as Now Practiced in India, Egypt, Italy, etc.* (1875). The report enumerated the "lessons of valuable experience" that India could teach California. With the aid of India's surveyor general, Davidson developed a comparative inquiry into California's irrigation future.[86] Congress received Davidson's ideas with reserve. Yet a strange confluence of events stimulated by the mining industry catapulted Davidson's ideas into the center of the political arena.

Davidson's report focused on Indian irrigation methods' relevance to California. He first stressed the great need for irrigation, especially west of the one-hundredth meridian. He recommended tapping into the Sierra Nevada and Rocky Mountains and developing irrigation by gravitation, as in India. An irrigation bureaucracy modeled after India's Irrigation Department would empower officers to govern the daily workings of canals, arbitrate conflicts, and ensure the efficient use of water. Modern technology posed no problem. "I have seen no difficulties of engineering," Davidson concluded, "which cannot be readily and successfully overcome by our engineers."[87]

Davidson's report was most radical for its promotion of the Northern India Canal and Drainage Act. "The government of India," he wrote, "exercises the absolute right to use and control the waters of all streams and lakes, without compensation." But it was "utterly impossible," he mourned, "that such conditions could exist in the United States except where the Government enters new territory."[88] Yet irrigation advocates argued that California's Great Valley was new territory to be publicly possessed, despite private owners' claims. Davidson hoped that policy makers would apply the same laws that prevailed in the Punjab to the valley and develop it "before the country is densely populated, and before conflicting [water] rights are too great." He stressed that the full exploitation of California's great resources depended fully on the adoption of a similar act.[89]

India's example had "double weight in California," Davidson warned, "because we are commencing without profiting by the dearly bought experience of others."[90] Davidson issued his advice too late. Farmers and ranchers such as Miller and Haggin had already monopolized vast stretches of land in the Great Valley. They also bought and sold water in exactly the way Davidson had recommended against—purchasing it as a separate commodity that could then be used against potential buyers and other water users. Davidson wished to apply the riparian principle to both mining and irrigation for precisely the reason that mining threatened valuable irrigable lands.[91]

In the 1860s and 1870s, hydraulic mining began to threaten the farms that had sprouted up in the Sacramento Valley in the 1850s. Hoses blasting water at one hundred miles per hour tore down entire mountainsides, depositing enormous quantities of mud, sand, and gravel into the Sierra rivers' canyons. This debris washed downstream in spring floods, silting the Sacramento, Feather, and Yuba Rivers and obstructing river navigation. Mud clogged more than thirty thousand acres of farmland in the Sacramento Valley. In response, residents in towns along the river built miles of levees, but property values continued to decline.[92]

Sacramento Valley farmers began to wage war against the most destructive corporations, including the North Bloomfield Gravel Mining Company, in the mid-1870s. When debris almost buried the towns of Marysville and Yuba City in 1875, farmers and townsmen formed the Anti-Debris Association. This diverse group believed that the public good had fallen to the wayside as hydraulic mining corporations continued to threaten valuable farmland. California's agricultural future, farmers argued, required that the actions of the great mining corporations be checked. Yet these companies owned many prior land and water rights and had begun their operations before farming took root in the Sacramento Valley.[93]

The battle between the Miners' Association, represented by North Bloomfield Gravel Mining, and the Anti-Debris Association, composed of residents in Sacramento Valley counties and farming towns, dominated the courts for years. Questions about California's future put the State Assembly into utter disarray: How could the Sacramento and its tributaries be reclaimed and controlled? Could the state do so alone, or must the federal government intervene? How could the government revive the mining industry while safeguarding the farmers? The hydraulic mining controversy forced Californians to grapple for the first time with the related problems of valley drainage, flood control, river navigation, reclamation, and water law.[94]

As the hydraulic mining controversy reached appalling proportions, the California legislature invited Davidson to present his views on irrigation to the State Assembly. In January 1878, Davidson addressed an expectant House, whose legislators knew little about the benefits of planned irrigation. Using the hydraulic mining controversy as the most blatant example of laissez-faire gone awry, Davidson focused on the dire need for the state to oversee irrigation development. Irrigation, he argued, meant "life or death to our advance and prosperity." Without irrigation, "the centre of gravitation of the population of the United States will not move westward of the Mississippi."[95] The hydraulic mining industry highlighted the monopolistic practices of private mining, irrigation, and canal companies, which thrived at the small farmer's expense. Davidson proposed that the state condemn and redistribute all privately held waters on the grounds of "great public necessity," as he believed the British had done five years earlier.[96] He recommended that the state first sort out its water laws, then "reconstruct the whole fabric, looking to the absolute rights of all, under the most intelligent views of a commission of Arbitrators composed of Engineers & Statesmen; and representatives of the great interests involved." Davidson's amorphous bureaucracy would invite the expert engineer, called the "deputy God" in India, to oversee irrigation development throughout the state.[97]

The California Assembly neither jumped into action nor circulated Davidson's reports.[98] Nevertheless, Davidson stepped into an exceptionally favorable political climate for irrigation. His views on water development, while radical for even the staunchest of irrigation advocates, met with general approval from opponents of hydraulic mining. The mining controversy, combined with the drought that had plagued California since 1876, catapulted Governor Will Irwin's legislature into action. Most of the Grangers and anti-Chinese Vigilantes who dominated that session agreed that the state needed an administrative body to address recurring drought, floods, and the debris question; to update California's water laws; and to draw comprehensive irrigation guidelines.[99]

Davidson's views on the government's duties bolstered sentiment already brewing in the California legislature. In January 1878, the State Assembly invited B. S. Alexander to give expert advice on the debris, flood control, and irrigation problems. In accordance with the recommendations of his 1873 commission, Alexander proposed that a centralized body of engineers begin by conducting a thorough investigation related to the concerns of all factions.[100]

The following February, mining debris again choked the Sacramento River and flooded its banks. In May, the state legislature passed a compromise measure affecting stockmen, irrigators, and hydraulic miners. Senator Creed Haymond wrote the first version of bill, and Mendell modified it. The bill created the office of state engineer. The legislature appropriated the enormous sum of one hundred thousand dollars "to provide a system of irrigation, promote rapid drainage and improve the navigation of the Sacramento and San Joaquin Rivers." Irwin then appointed his friend, William Hammond Hall, to the head post of state engineer, with a salary that matched the governor's. Alexander and Mendell became Hall's personal advisers, although Alexander died shortly thereafter.[101]

Hall was already well connected to California politics. In 1872, Ralston had asked Alexander for advice on the engineering plans for the San Joaquin and King's River Canal Company, and Alexander in turn had requested that Hall survey the situation in the valley. When the San Joaquin and King's River Canal Company went bankrupt in 1876, the Bank of California and the Nevada Bank sent Hall to take charge of the old company's tens of thousands of acres of land and its several large canals. The same year, a number of large landowners farther south whose holdings would have been irrigated under Ralston's original plan secured a law creating the West Side Irrigation District. The governor named Hall as chief engineer. From there, he rose to state engineer.[102]

Hall represented one of a growing cadre of engineers who in the late nineteenth century helped make California a hub for the global circulation of irrigation and mining technology as well as ideas about frontier expansion, society, and the role of government in economic development. Together, California engineers' technologies and ideas formed templates of economic growth for other parts of the world. Some of their models were indigenous to California, evolving out of its laissez-faire culture. Others were imported from abroad, reformulated to fit California's conditions, and then reexported to other countries. We now turn to a more careful examination of California, where these various models of development, from state oversight over irrigation works to planned irrigation colonies and federal reclamation, played out to varying degrees of success.

### CHAPTER 2

# A Great Mission for the Race

### LESSONS AND EXPERIENCES FROM CALIFORNIA

In his position as state engineer, William Hammond Hall embodied a grow-ing reliance on technology and rhetoric shared by many Californians. In-deed, California engineers' models of irrigation shared a host of traits, not all of them technical. Many late-nineteenth-century engineers and policy makers used Jeffersonian rhetoric about grassroots democracy, local con-trol, free markets, social mobility, and faith in the common man to gain popular support for projects that intended to reorder the landscape and society. They imbued these ideals with the modern values of scientific and expert management. As Californians transitioned from a frontier to indus-trial society, engineers' experiments in redefining the state's, market's, and society's roles in economic growth shaped frontier development around the world.

One international irrigation model emerged from Hall's State Engineer-ing Office in the 1870s and 1880s, but although Hall reworked aspects of the Indian model, he ultimately failed to build a statewide irrigation policy and system. An alternate form of development in the 1890s involved Southern California's planned irrigation colonies, which meshed common ideals and ownership of irrigation work with private landownership. A third model of economic development, federal reclamation, once again sought to use government power to encourage agricultural growth. But until the federal Central Valley Water Project of the 1930s, none of California's irrigation models reconciled the market, private enterprise, and government over-sight in socially, politically, or economically sustainable ways.

California engineers nonetheless exported these models of economic growth around the globe, implementing them quickly if not always successfully. Australians adopted California's water laws and irrigation colonies. Californians' mining and irrigation expertise as well as ideas about free markets and political liberty went to South Africa. Hawaii and Palestine drew on California's agricultural settlement plans. The spread of Californians' models, a reflection of how American technology, ideas, and personalities interacted with foreign governments and cultures, helped create global markets and shared technical expertise. It also furthered a "great mission" for white settler societies around the world, where it produced extremely different lessons from those learned in California.

### Frontier Development and State Intervention in California

Many late-nineteenth-century Californians' vision of frontier progress rested on the belief in private enterprise, minimal state intervention, and local political control. Yet the hydraulic mining controversy and the state's haphazard irrigation growth illustrated the negative effects of these ideals on agriculture. Unregulated hydraulic mining, coupled with an 1876–79 drought, devastated Northern California's wheat fields and orchards. These problems indicated the need for a water system that would provide authorities with the means to reconcile mining and farming interests, clear mining debris from rivers, design new water laws, untangle overlapping water claims, improve navigable waterways, and expand agricultural lands through irrigation.

As state engineer, Hall addressed many of these issues. He tried to use government power to catalog the state's water resources and mitigate damage from the hydraulic mines and drew on India's state-oriented model for his ambitious drainage plans. Yet California's antiquated water laws; recurring conflicts among farmers, ranchers, and miners; and patterns of land and water monopoly made the direct application of India's water regime impossible. Hall thus combined state control over water with private control over irrigation works. Yet his attempts to halt mining damage and bolster irrigation failed; in the end, the shortcomings of state planning reflected the continuing supremacy of California's regional and private interests.

Hall was born in 1846 in Hagerstown, Maryland, into an influential family of British lineage. He proudly identified some of California's most colorful figures as family: Texas Ranger captain John Hays, who served as San Francisco's sheriff; John Hays Hammond, a famous mining engineer; and Major Richard Pindell Hammond Jr., the state's surveyor general. Hall's family settled in Stockton when he was seven. As a young boy, Hall had no intention of becoming an engineer; he only reluctantly gave up his horse and guns for surveying instruments. His family wanted him to attend West Point to study engineering, but the outbreak of the Civil War curtailed these plans. Instead, unlike many of his future colleagues, Hall received his engineering education "by bits or small chunks and irregularly." This self-education left him slightly arrogant and opportunistic.[1]

Vast opportunities for engineers arose in the mid–nineteenth century, as U.S. settlement advanced west of the Mississippi Valley, the Mexican War produced a vast new American territory, California experienced its gold rush, and officials planned the construction of a transcontinental railroad. The end of the Civil War in April 1865 initiated a flurry of federal engineering activity on the Pacific Coast. New agencies, including the U.S. Board of Military Engineers and the U.S. Geological Survey, led by John Wesley Powell, began to link the acquisition of scientific knowledge to the nation's commercial future.

These early surveys gave Hall valuable experience. In 1865, he joined the U.S. Corps of Engineers in Oregon as a draftsman. Between 1866 and 1870, he worked as a field engineer for the U.S. Board of Engineers for the Pacific Coast under General B. S. Alexander, his lifetime mentor. With Alexander, Hall surveyed locations for military posts and harbors from Southern California to Puget Sound. In 1871, after a topographic survey of San Francisco's Golden Gate Park, he became the park's superintendent and chief engineer, a post he held for five years and in which he worked with John McLaren and Frederick Law Olmsted, the architect of New York City's Central Park, to design Golden Gate Park's landscape.[2]

Though he was not the shrewdest of politicians, Hall made alliances during this time that cemented his reputation as one of the state's leading engineers. In 1876, Hall left Golden Gate Park to head the State Irrigation and River Commission (the West Side Irrigation Commission). William Ralston's bankruptcy and ranchers Henry Miller and Charles Lux's subsequent acquisition of the San Joaquin and King's River Canal and Irrigation Company's properties prompted wheat farmers to ask the government for aid in building a canal. Governor Will Irwin named Hall chief engineer and Alexander consulting engineer for California's first river and water supply examination. Hall surveyed several thousand acres of land and reported on the feasibility of designing a canal that would draw water from Tulare Lake, a plan that resembled Robert M. Brereton's original project. At the same time, the Bank of California and the Nevada Bank, whose presidents had sat on the San Francisco Park Commission board, hired Hall to take charge of the defunct San Joaquin and King's River Canal and Irrigation Company's properties.[3]

Hall's professional reputation spread quickly. In the summer of 1877, Creed Haymond, a legislator representing Sacramento County in the State Senate and chair of the Senate's Committee on Irrigation, approached Hall and suggested that the two men write a bill promoting irrigation. A few weeks later, Haymond showed Hall the draft of his proposal. It embodied Haymond's ideas about a hydraulic mining and river improvement investigation and Hall's plan for a statewide irrigation system. The bill became the law that created the State Engineering Office. "Nobody," Hall admitted, "except Senator Haymond, Ex-Governor [John G. Downey] and Governor Irwin knew that I had been a party to the move."[4]

The State Engineering Office was a political compromise. It not only served those who saw the need for state intervention in accordance with the Indian model but also showed a lack of commitment to a thorough overhaul of existing water policies. Hall pessimistically believed that the office was simply "a hoped-for arbitrator" among several small partisan groups and claimed in later testimony before the U.S. Senate that Irwin had created the office as a compromise measure between stockmen and irrigators.[5]

Political controversies hampered the State Engineering Office's work from the start. California's antiquated water laws posed the first obstacle. Under indefinite state law and custom, prior appropriators took as much water as they chose. The lower bank and adjacent riparian owners claimed water under a different English common law. Hall thought that both positions were unreasonable. Like George Davidson, Hall especially opposed the doctrine of prior appropriation. In his mind, the legislature had created the State Engineering Office with the unrealistic expectation that "by some hocus-pocus or feat of 'science'" both interests could be appeased.[6] Instead, the battle between water appropriators and riparian rights as well as regional conflicts came to a head during Hall's tenure.

Northern California's antidebris and hydraulic mining factions also clashed. Miners claimed that runoff from plowed lands harmed the rivers as much as the tailings from mining. They believed that the mining debris, unlike agricultural sediment, could be kept out of the streams.[7] The antidebris faction, mostly farm and town property owners bordering the polluted streams, disagreed. Yet the conflicts between miners and farmers in the Sacramento Valley were more complicated than Hall initially thought, as some mining and agricultural interests overlapped.[8]

Central California experienced parts of the north and south's problems. According to Hall, the swampland irrigators tampered with the public drainage ways and navigable streams by constructing private levees; for their part, farmers and ranchers denied the necessity for state control of the levee systems but wanted the state and federal government to dredge rivers and construct relief canals. Yet most of Central California's inhabitants shared northern farmers' general belief that hydraulic mining should be stopped.[9]

Hall realized that the problems of irrigation, mining debris, drainage, and navigation were interrelated. One could not be solved without addressing all. Facing enormous political and regional blocs, Hall could devise few solutions that all groups would find palatable—his power, unlike that of British engineers in India, was limited from the start.

Hall began his work by continuing parts of the survey that Davidson, George Mendell, and Alexander had begun. In 1878, Hall dispatched five survey parties to examine the extent of California's problems. Alexander, the ranking U.S. Engineer officer on the Pacific Coast, and Mendell, an engineer and colonel in the Army Corps of Engineers in charge of the government river and harbor works in California, headed the hydrographic surveys. Alexander and Mendell measured rainfall, snowfall, stream flow, drainage, ground and surface storage, and flooding. They also gauged the mining debris deposited by the Yuba, Bear, Cosumnes, American, and Feather Rivers and tributaries and mapped part of the Sacramento River. They then charted 170 miles of the San Joaquin River and its tributaries, focusing on the irrigation possibilities in Southern California counties.[10]

Hall presented the survey's findings to the legislature in 1880, focusing on hydraulic mining debris in the Sacramento Valley and the flooding of the Sacramento River. The combination of water and mining debris, composed of boulders, stones, gravel, fine sand, and powdery mineral and soil particles, formed a thick muck that choked the entire river system and devastated orchards and towns. Mountain canyons held enormous amounts of debris. Tailings had made Steamboat Slough, an important navigational branch of the Sacramento River, nearly impassable.

Hydraulic mining, Hall reported, had wrought a great change in the landscape. Agricultural lands bordering the Feather, Yuba, and Bear Rivers had been very fertile prior to the debris. Before the era of hydraulic mining, floods occasionally surged through the rivers but receded rapidly and rarely affected crops. During the height of hydraulic mining, prosperous homes, orchards, and fields experienced more frequent and devastating floods. Sand filled the rivers' natural channels, and levees spilled over. In places, debris buried orchards, gardens, fields, dwellings, and landmarks.[11]

In 1848, the Feather River's waters had been clear and its banks high and clearly defined. Thirty years later, debris had raised the water level, destroyed the shoal bars, and widened the Feather's course. Hall concluded

GOLDEN GATE AND GOLDEN FEATHER MINING CLAIMS

*Golden Gate and Golden Feather mining claims along the Feather River, Butte County, California. Courtesy of the Bancroft Library, University of California, Berkeley*

that tailings had drowned some 1,500 acres of prime agricultural land in the river's fertile basin. Land planted with corn and potatoes in the spring of 1878 now lay six to eight feet underwater. The value of land had decreased from $100 to $10 an acre. Near Yuba City, high water had destroyed more than ten thousand orchard trees, costing the community $63,000. The Yuba and Bear Rivers suffered equally. Assistant state engineer Marsden Manson warned that floods could sweep two hundred thousand cubic yards of nearby mining debris into towns and outlying agricultural lands at any moment. Debris had buried more than 7,600 acres of orchards near Marysville. Only treetops protruded above the sand. Hall noted that the debris from the North Bloomfield mines alone created scummy ponds thirty feet deep that backed up for miles. Slime covered more than 8,800 acres of prime agricultural land along the Bear River. Hall estimated that hydraulic mining had destroyed more than 43,500 acres in the Sacramento Valley, leading to a depreciation in value of $2,597,635 and costing the state $7,143 in taxes annually.[12]

*Malakoff Diggings: hydraulic mining at North Bloomfield, Nevada County, California.*
*Courtesy of the Bancroft Library, University of California, Berkeley*

Hall presented his findings to the legislature in March 1880. Soon thereafter, the large mining companies created a bill for a special state drainage and debris commission. The Drainage Act passed the State Assembly
by a vote of forty-three to thirty-six and the Senate by twenty-one to sixteen, with obvious sectional alignments. Every county south of Merced, including the San Joaquin Valley, opposed the bill. Counties in the Sierra and
Lower Sacramento Valley, which would benefit most from the new commission's work, solidly favored the measure.[13]

The Drainage Act created the Board of Drainage Commissioners, which
included the governor, the surveyor general, the state engineer, and other
experts, including Davidson, James B. Eads, and Mendell.[14] The legislature
first charged the commissioners with dividing the state into drainage districts, as British engineers had done in India's northwestern region.[15] Each
district would act as a unit in promoting drainage in its territory and implementing Hall's flood control schemes. The commission would then design

rivers as flood-carrying and navigable channels to straighten and align some of the banks. Flood escape weirs would prevent a repeat of the damage caused by the floods of 1853, 1862, and 1868.[16] Hall understood the natural cycles of rivers as well as any scientist at the time. But because in some areas the damage caused by hydraulic mining only hastened the natural flooding and shifting of rivers, some of the effects of the debris could not be quantified.

Hall also drew on the largest mining companies' suggestion that impoundment dams be built to trap mining debris and store water for irrigation. Debris dams would serve the general welfare by restoring the prosperity of both the hydraulic mines and valuable agricultural land, protect the cities of Marysville and Sacramento from serious flooding, and preserve streams from irreparable injury.[17]

Various groups opposed the Drainage Act from the start. Many California residents feared unlimited taxation. The state raised revenue for the act by levying a general tax of .05 percent on all property in the state and imposed a district tax. The state further taxed all reclaimed land and required owners of all hydraulic mines to pay an annual assessment tax of .5 cents per each miners' inch of water used.[18] State officials argued that because mining companies had created the problem, they should pay the full cost of protecting the Sacramento Valley from further damage. Southern Californians complained that the entire state would unjustly bear the cost of the work for the Sierra districts. Others maintained that the whole scheme was unconstitutional because it bestowed governmental powers on a special commission. To make matters worse, few people believed that Hall's proposed debris dams would work. Even Hall had reservations about the Drainage Act. Although he never publicly voiced his opinions, he believed that the act gave too much power to the state and that the federal government should have shouldered half the cost.[19]

The legislature asked Hall to design the debris dams in April 1880. That fall, eight hundred workers began to construct two restraining dams on the Yuba and Bear Rivers. Given the dire circumstances, the dams succeeded as a short-term stopgap measure; in the long term, however, they formed no component of the water system Hall imagined.[20] Eads convinced Hall to forgo earth-fill or masonry structures for cheaper dams made of wire, brush, willow, and tree layers, with bags of sand weighted on top. These brush dams kept water spread over a wide area and did not fully block stream flow. Still, engineers from around the state praised the structures. The Yuba dam had a brush and sapling embankment eighty-nine hundred feet long, and the Bear River had a similar, slightly smaller structure. By July 1881, the

Yuba River brush dam had caught almost five million cubic yards of debris, although one-seventh of the structure had been washed away.[21]

Hall secretly expressed his disappointment with the brush dams. Yet rather than admit that his mastery over nature and circumstance had not been as complete as he wished, Hall blamed his consulting engineers for all mistakes in cost, material, and planning.[22]

Various factions that had tried to repeal the Drainage Act from the start compounded Hall's accusations. In March 1881, the Supreme Court of California rewarded Hall's opponents by repealing the act on a technicality, forcing the State Engineering Office to halt all planned debris works. The state had spent more than five hundred thousand dollars but had little to show for its efforts. The Debris Commission had created only one drainage district in the Sacramento Valley. The dams, which required maintenance and had been damaged by a rainy winter, went to ruin. In October 1881, a fire destroyed the remainder of the Yuba River dam.[23]

The repeal showed the extent to which the state lacked power to resolve the competing interests of miners and farmers, northerners and southerners as well as the State Engineering Office's inability to reach consensus from within and to design long-range solutions to resource problems. Most of all, the act's demise illustrated the supremacy of private interest groups over government planning.

Although the Drainage Act produced no lasting results, it did signify a turning point in state resource planning. For the first time, the state committed funds to water development and recognized the need for centralized planning. The measure resulted in the first large-scale effort to control the Sacramento River and its tributaries with the understanding that different groups of people could affect the entire ecosystem. The Drainage Act was also revolutionary in that it used government as a tool to try to find consensus among competing interest groups in the name of the greater general welfare.[24]

The problems that weakened the power of the Debris Commission also hindered the State Engineering Office's sweeping plans for irrigation. Vested private interests, the widespread belief in individual enterprise, and a lack of government funds curtailed all of Hall's projects. So did his radical proposals, in particular his call (echoing Davidson) for state ownership of waterways.[25] The Northern India Canal and Drainage Act, which had in theory granted the British government the supreme right to claim and use water in northern India, provided Hall with one model for California irrigation. Like Davidson, Hall recommended that Californians develop their water system on the principle of inalienable public ownership of waterways and establish regulations that would prevent water monopoly. Since the ownership of water had given rise to these serious conflicts, changing the type of ownership might fix them. The "sooner it is done," Hall argued before the California State Legislature in 1881, "the better for all concerned."[26]

Yet Hall did not look exclusively to British India. He realized that the principles behind India's centralized hydraulic regime would have to be greatly modified to fit California's open economy and culture. Hall thus found more relevance in the irrigation regimes of France, Spain, and Italy, which had not been subject to private appropriation since feudal times. Water remained the common property of all people. Clear and enforceable water laws short-circuited legal conflicts. Despite this measure of state authority, however, local associations and small farmers, particularly in Spain, controlled most of the irrigated regions and works, thereby providing Hall with a model of state-monitored private enterprise.[27]

Public ownership of irrigation and flood control works, as in northern India, was the logical corollary to state ownership of water. The British government controlled both rivers and irrigation works, sending out hordes of engineer-bureaucrats to manage local systems. Yet in Hall's mind, public control of waterways did not mean construction of public works, interference with private industry, or the management of bureaucrats.[28] Rather, Hall believed in state-guided private enterprise, a popular point of view that stemmed from a fear of an intrusive government. Hall maintained that the state bred corrupt politicians, promoted waste in resource use, and disdained local control.[29] His reports thus advocated the creation of a system of individual property rights based on irrigation districts that conformed to natural drainage basins. The irrigation districts would act as state-monitored corporations by doling out water rights and monitoring development, but the construction of irrigation works would remain in private hands.

Hall proposed irrigation bills to the state legislature in both 1880 and 1881. The first recommended the creation of a state board of water commissioners consisting of the governor, surveyor general, and state engineer—not quite the bureaucracy that Davidson had witnessed in India. The board would ascertain the extent of claims to water for agriculture, mining, and manufacturing purposes. It would also establish standards for measuring water quality, the volume of diversions, and the need for different crops. The second bill would have created local administrations of three to five counties each. These irrigation and drainage districts would oversee the daily administrative tasks of irrigation, including monitoring and regulating water diversions and constructing their waterworks.[30]

Hall's plans for irrigation met the same fate as his debris dams. The legislature rejected both of his irrigation bills, thereby missing an opportunity to have the most advanced water laws in the West, an honor that went to Wyoming under state engineer Elwood Mead. Hall's proposals that the state govern all water, even if the actual dams and canals remained in private hands, struck at almost every water user in California. Mining blocs attacked Hall's irrigation bills. Farmers in Southern California and the San Joaquin Valley opposed all measures they thought would aid Northern California water users. Simply put, long-range state planning lacked widespread support in the 1880s.[31]

The end of the debris dams, the repeal of the Drainage Act, and Hall's forsaken bills discredited the State Engineering Office. The 1881 legislature appropriated less than half as much money for the irrigation surveys, curtailing Hall's grandiose schemes. For example, he published only two of his planned seven books on the social, political, legal, physical, and technical aspects of irrigation.[32] The first volume examined irrigation and water law in France, Italy, and Spain, a topic the public perceived as too esoteric. The second surveyed irrigation in Southern California. Both circulated poorly. Hall and his assistant state engineer, James D. Schuyler, also began *Water and Land*, a journal devoted to the development of resources of the Far West. It survived only two issues. Budget constraints all but eliminated Hall's hydrographic survey work, and funding for the office kept shrinking.[33]

Hall's popular support dwindled as well. Many legislators as well as Governor Robert Waterman believed that Hall should be spending more time clearing rivers than dreaming. Hall, who blamed the debris dam failure on Mendell and Alexander, also alienated his colleagues. The public viewed Hall's theories about the value of government regulation in a laissez-faire economy as too impractical. Finally, the heavy rainfall of 1880 and 1881 made the drought years of the mid-1870s and consequently the need for irrigation seem like a faint memory.[34]

By 1881, Hall was so disillusioned with public life that he threatened to resign as state engineer. His plan to systematize water planning involved too many compromises and sacrifices. He was not, however, wholly responsible for his failures. The dwindling popularity and power of the State Engineering Office reflected the larger political climate. Indeed, three important pieces of legislation revealed the state's ambiguous and even contradictory relationship to society. The first illustrated the government's power to curb private enterprise. The other two showed the opposite—the persistence of private and local control over long-range, state planning.

The first piece of legislation struck at the hydraulic mining industry. For years, the Anti-Debris Association had waged a fierce battle against the Hydraulic Miners' Association, a group comprised of the big mining companies. In *Edwards Woodruff v. North Bloomfield Gravel Mining Co., et al.* (1884), Judge Lorenzo Sawyer ruled that hydraulic mining constituted a general and destructive public and private nuisance that must be halted. The suit covered all mines depositing debris into the Yuba and its tributaries and expressed Californians' growing concern for the destructive tendencies of unregulated enterprises.[35]

The second landmark case addressed the conflict between riparian rights and prior appropriation. What both Hall and Davidson had feared came to pass: a showdown between riparian owner Miller and his southern neighbor along the Kern River, James Ben Ali Haggin, who owned 413,000 acres of land and immense quantities of water he had claimed under prior appropriation. His claims diminished the flow of the Kern River, to the detriment of downstream riparian landowners Miller and Lux. The battle between Lux and Haggin dragged on between 1879 and 1886, and the state supreme court eventually sided with Miller, Lux, and riparian rights. Water users considered *Lux v. Haggin* a hodgepodge victory. Rather than creating transparent laws, the court invented a legal hybrid in 1884 known as the California Doctrine that ultimately favored both systems of water use. Far from solving the key water rights issues, the decision confused them. It also failed to alter the state's concentrated pattern of land and water rights. But by that time, the two warring water lords had combined their empires.[36]

The *Lux v. Haggin* decision paved the way for another piece of legislation, the Wright Act of 1887, which created irrigation districts. In 1882, the state legislature had rejected Hall's plan to create irrigation districts whose water supplies would be guaranteed by the state. The 1887 act authorized the residents of defined areas to form local districts, issue bonds and raise revenue, purchase (and condemn) land and water rights, and distribute water through private companies. In theory, all rights belonged to the district, not to individual water users. By 1895, forty-nine districts covered 2 million acres of land, 2 percent of California's area, but most operated inefficiently. Although irrigation was extended to 1.5 million more of the state's 28 million acres of farmland, much of that growth occurred on large estates. Periodic floods, the 1889 drought, and the 1893 depression worsened conditions. Overall, the Wright Act promoted uncoordinated, local development. Hall, who had publicly denounced parts of the measure, argued that Californians needed a more centralized solution. But the tide had turned against state planning.[37]

The fate of the State Engineering Office reflected this sentiment. In

March 1888, Hall was formally investigated, though acquitted, for the unauthorized spending of state funds. He had allocated twenty thousand dollars to future costs rather than have the money revert to the general state fund. Although he was cleared of personal dishonesty, he resigned as state engineer in January 1889. Soon thereafter, the legislature abolished the office and turned its records over to the state mineralogist.[38]

During his eleven years as state engineer, Hall, whom Pisani describes as both "visionary and impractical," created precious little irrigation policy. Nevertheless, the State Engineering Office trained many of the state's most prominent hydraulic engineers, including Schuyler, Manson, and Carl E. Grunsky, who went on to contribute significantly to irrigation development both in California and in the world at large. The office also gathered impressive statistics on California's physical geography, irrigation, and water policy. Hall's belief that long-range state planning could produce a society of small farmers even foreshadowed some of the principles behind later water projects, including the 1930s Central Valley Project. Finally, many of the conclusions in Hall's reports found popular support around the world. Although he failed to implement his model of development in California, he transported parts of it to other frontier societies.[39]

## Southern California's Colony Builders: Ontario, the Model Colony

The abbreviated life of the State Engineering Office revealed the government's shortcomings in irrigation planning. But as Hall battled the state legislature, Southern California offered a valuable model to the rest of the state and the world: the irrigation colony, part of Southern California's first agricultural land boom. Many of these settlements offered an alternative form of development to the irrigation district and individual farm. The Chaffey brothers' famous Ontario settlement, the "model colony," showed how private developers could create modern rural communities by subdividing large estates, designing community-owned water systems, and building cooperative marketing and production facilities. Yet despite its planners' rhetoric about social mobility and grassroots democracy, Ontario embraced an elite class of farmers. Nor could the colony, in many ways the exception to similarly planned settlements, be replicated successfully elsewhere.[40]

Ontario owed much to its predecessors: the Mormon settlement at San Bernardino; the German colony at Anaheim; Riverside; and Etiwanda. In 1851, the elders of the Mormon Church created a template for community farming when they purchased the San Bernardino Rancho. Colonists planted two thousand acres of grain and a forty-acre vineyard, and the Santa Ana River provided the colony with water. This settlement was the first of a string of towns along the "Mormon corridor" linking Salt Lake City and Southern California. Yet only six years after the colony's founding, the settlers returned to Utah.[41]

Also in 1857, fifty German tradesmen formed a successful irrigation colony at Anaheim, forty miles southeast of Los Angeles near the Santa Ana River. The three hundred colonists cleared and divided the cactus-ridden land. The founders dug a five-mile ditch from the Santa Ana River and ran lateral canals to the different properties. They sidestepped the complicated issue of water rights and distribution by devising a mutual water company, a private corporation that revolutionized water distribution by placing control of the company and its works in the hands of the property owners through stock. Anaheim provided a model for the mutual water companies and principle of local control over water that prevailed in parts of Southern California, including Ontario, by the end of the 1880s.[42]

The Riverside Colony followed Anaheim's example. In 1870, a group from Tennessee set out for Southern California, purchased land from a Mexican rancho owner, and prepared small farms. The 1876 completion of the Southern Pacific's line into Los Angeles touched off an agricultural colony boom in Southern California that lasted well into the following decade. Riverside's homesite prices rapidly increased from forty dollars to five hundred dollars an acre, and orange groves from one thousand to two thousand dollars per acre. By 1888, Riverside boasted the region's most advanced irrigation system, with canals and ditches spread over more than eighty miles and a reservoir and domestic pipeline system. Settlers also adopted a cooperative marketing system and fruit exchange.[43] As the pioneer orange colony prospered and gained fame, California irrigation advocate William E. Smythe remarked that irrigation worked "a revolution in the character of the people and country" by putting a class of landed proprietors at the base of society.[44] Small, irrigated farms—along with fine stores, churches, hotels, public halls, a clubhouse, and a library—created what he described as "the highest standard of civilization."[45]

While colony building was often a highly speculative venture, these settlements created a new standard of rural living. Etiwanda and Ontario set even higher standards. The colonies' founders, George and William Benjamin "Ben" Chaffey, were born in Ontario, Canada, in 1848 and 1856, respectively. Their father, George Chaffey Sr., Ben, and a sister immigrated to Southern California in 1878, settling in Riverside. Ben soon earned a reputation as a horticulturist. When George visited his family in the winter of 1880, he stayed to help Ben design Etiwanda, the first of their two irrigation colonies in Southern California.

Named after an Indian chief who had befriended the Chaffeys in Canada, Etiwanda was Ontario's trial run. In 1881, the Chaffey brothers purchased 560 acres from J. S. Garcia. The Chaffeys liked the quality of the land, mild climate, water supply flowing down three small neighboring canyons, and the water rights that were already attached to the property. The tract lay forty-eight miles east of Los Angeles in the San Bernardino Valley, the largest alluvial valley at the foot of the Cucamonga Mountains, directly north of Riverside, on the main lines of the Pacific-Electric and Santa Fe Railroads and within two miles of the Southern Pacific. The Santa Ana River ran through the valley's southwest corner.[46]

After negotiating with other private owners, the Chaffeys added fifteen hundred more acres to Etiwanda and incorporated a company, Chaffey Brothers, to manage their holdings. Etiwanda's technical sophistication distinguished it from other irrigation colonies from the start. The Chaffeys divided the land into ten-acre blocks and designed a gravity irrigation system that carried water down from the mountain streams in wood flumes to a large reservoir. From there, concrete pipes transported water underground to each plot of land. The Chaffeys' cement pipeline system, which decreased the loss of water through evaporation and leakage, revolutionized irrigation systems in California.[47]

The Chaffeys advanced Anaheim's model water system and created a new prototype for Southern California's irrigation companies. George firmly believed that the settlers had a right to own all of the water available for irrigation in a manner proportional to their individual holdings. Chaffey Brothers transferred to land purchasers one share of water stock with each acre sold, limiting the stock in the Etiwanda Water Company to ten shares for every inch of water measured in a designated weir each year.[48]

The community-owned mutual water company differed from the irrigation districts created by the Wright Act. Because the largest landowners held a correspondingly large amount of stock, they exerted greater influence in dictating company policy. The mutual water company also guaranteed that settlers would not buy land without water.[49]

The Chaffeys began to advertise Etiwanda land in 1882. Hundreds of settlers poured in from Canada and the eastern United States. The brothers sold fourteen hundred acres of land in eight months. By 1885, more than six thousand acres of land was under irrigated cultivation. Marketing facilities improved within a few years, and settlers formed a local Growers' Cooperative Association. Grapes, oranges, and lemons thrived and brought great prosperity to the colony.[50]

The Chaffeys also pioneered electrical (arc and incandescent) lighting. George was the first engineer in the American West to file claims on mountain streams for electric current; Etiwanda was the first place on the Pacific Coast to use this energy. George organized the Los Angeles Electric Company, which gave the growing city its first light and power plant and made it one of the first cities in the United States to be lit exclusively by electricity.[51]

Etiwanda's success inspired the Chaffeys' second colony, Ontario, the largest reclamation project ever undertaken by two individuals. In 1882, Chaffey Brothers purchased the remainder of Garcia's land, about sixty-two hundred acres, with some water rights, for sixty thousand dollars. The Chaffeys added railroad and government land to this purchase and began to build on the site at the foot of the juncture between the San Gabriel and San Bernardino Mountains, on the Cucamonga plains in the eastern part of the foothill belt. Located six miles west of Etiwanda and thirty-eight miles east of Los Angeles, Ontario runs seven miles north to south and varies from one to three miles wide. The valley slopes southward, dropping from an altitude of twenty-five hundred feet above sea level to nine hundred feet over its eight-mile length. The San Antonio River furnishes water. Five alluvial fans merge to form the plains below the canyons of five main streams, and the upland loam soils are well suited for citrus crops.[52]

Ontario set a new standard for rural development. The Chaffeys laid out the main thoroughfare, Euclid Avenue, two hundred feet wide and eight miles long. At the Southern Pacific Railway crossing, a majestic fountain proved that Ontario had water. Four rows of eucalyptus, magnolia, palm, and pepper trees shaded the main artery. The Chaffeys also built garden homes, villa sites, ten-acre fruit farms, and business sites.[53]

The Chaffeys advertised their settlement in typical boom-period fashion. Pamphlets, newspapers, and special excursions to the colony touted Ontario as the "model colony," a name coined by Victoria's Alfred Deakin in 1885. "The proposition is a safe one for you," one advertiser claimed, "and it is certainly in marked contrast with fake schemes so successfully worked by which the fakir gets your money before you see the worthless realty you have bought."[54] At the start of Southern California's great land boom in 1886, Ontario's real estate prices rose daily. By the spring of 1887, speculation was in full swing. Over the rest of that year, people bought extravagantly in hopes of selling out to someone else at a good profit. Ontario's land sales peaked in August 1887 and did not slow until the next year.[55] By 1888,

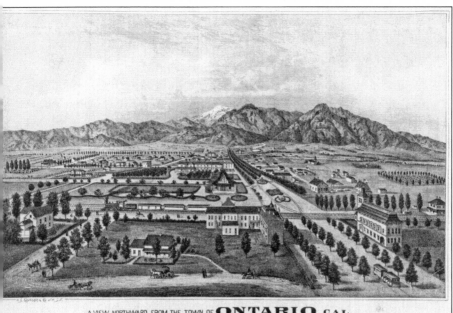

A VIEW NORTHWARD FROM THE TOWN OF ONTARIO CAL.
WITH SAN ANTONIO HEIGHTS AND, SIERRA MADRE MTS. SIX MILES DISTANT
ONTARIO LAND AND IMPROVEMENT Co. PROPRIETORS, C. FRANKISH MANAGER ONTARIO CAL.

*View northward from the town of Ontario, California. Courtesy of the Bancroft Library, University of California, Berkeley*

more than twenty-five hundred colonists from Los Angeles, San Francisco, New England, New York, Kansas, Missouri, Texas, and even London had settled into rural life.[56]

Ontario, however, required men with means. Few of the settlers fit the stereotype of the uneducated hayseed farmer in overalls. The *Los Angeles Times* boasted that Ontario drew "with irresistible force the moral, the intelligent, the educated and the refined," among them citizens of national reputation.[57] Indeed, the Chaffeys advertised Ontario not as a pioneer settlement but rather as a community of "prosperous city people" who enjoyed urban life "in miniature."[58] In "no other part of the civilized world," one irrigation booster exclaimed, "is there to be found more intelligent farmers than dwell amidst the orange groves of southern California."[59]

This urbane middle- and upper-class white population contrasted sharply with Ontario's workers. In the early 1890s, Chinese laborers established a camp a quarter of a mile from the colony. All of the men worked in the fields, and some of the women assisted with household chores. Mexican and Indian field hands replaced the Chinese laborers by the 1900s, replicating the pattern of ethnic labor found in the rest of California.[60]

The fruits of this labor allowed the Chaffeys to portray their colony as a "Modern Arcadia."[61] One Ontario settler, Judge Robert M. Widney, praised the desert's almost biblical transformation. "The land that was supposed to be worthless," he wrote, has "turned ... into gold."[62] A writer for the *Ontario Fruit Grower* marveled that "when last I saw this place it was part and parcel of the surrounding desert, while today it promises to be a veritable Garden of Eden."[63] This Eden provided a panacea for America's overcrowded cities but of course owed its existence to Los Angeles markets and the extensive rail networks that connected the colony to the rest of the nation. The transcontinental Santa Fe Railroad reached Ontario in early 1887, with the Southern Pacific and Salt Lake lines arriving in the early 1900s. Boosters even predicted that Ontario would benefit from the proposed Panama Canal.[64]

Ontario's most notable technical innovation was its world-class water system. By tapping the subterranean flow of the San Antonio Canyon, George Chaffey built Southern California's first tunnel for water.[65] Cement was manufactured for the pipes at Ontario; more than twenty-five miles of irrigation pipes were laid, and a dam was built in the canyon. These engineering works led settlers to believe in nature's endless bounty. "Droughts are impossible," one homeowner noted. "Water for irrigation and drinking comes from an inexhaustible source."[66]

In 1903, the Chaffeys completed Ontario's hydroelectric plant, one of the first in the nation to combine irrigation and hydroelectricity. Colonists owned both the mutual water company and the Ontario Power Company, a phenomenon the *Los Angeles Times* described as "the progress of Southern California."[67] In 1903, the U.S. government selected Ontario as the standard for American irrigation colonies. Federal engineers subsequently constructed a model of the irrigation and hydroelectric system to exhibit at St. Louis's 1904 world's fair. Not surprisingly, Ontario also created a prototype for municipal utility systems. In the 1890s, as San Francisco and Los Angeles politicians began to campaign for the public ownership of utilities, Ontario provided a model for such a system, especially after becoming a municipality in 1891. One booster noted that Ontario provided the "best examples of public ownership of what ordinarily become profitable private monopolies" in California.[68]

In 1886, the first year of Southern California's great land boom, the Chaffeys left Ontario to plan and run similar settlements in Australia. The Chaffeys subsequently sold about seventy-five hundred acres of land and waterworks at great profit. They conveyed their holdings, which included water rights and the Ontario Hotel and offices, to San Francisco and Los Angeles businessmen for $275,000.[69]

Although municipal incorporation brought new industries to Ontario, cooperative agriculture remained the core of the town's existence. The Chaffeys had started the colony's nursery in the 1880s with forty thousand peach seedlings, twenty thousand pear trees, fifteen thousand apple trees, one thousand pepper trees, and thousands of orange trees.[70] The growers organized into marketing cooperatives when the trees bore fruit. Although a few private companies flourished, such as Sunkist oranges, cooperatives took over most of the picking, packing, and shipping. Most operations, except for the picking by Chinese and Mexican laborers and final wrapping of each fruit, were mechanized in the thirteen packinghouses, four fruit-drying operations, large cannery, and plant for the manufacture of citric acid. By the end of the 1910s, the value of Ontario's land had risen so greatly that land with bearing orange groves cost at least one thousand dollars an acre.[71]

Ontario's social ideals matched the orderliness of the town, agricultural lands, and industries. Ontario was the first colony in Southern California if not the entire state to prohibit liquor traffic, an act designed to sidestep the social ravages caused by disreputable saloons and to attract families of good stock. "Home life is without evil influences," one settler noted.[72] Prohibition remained in place until the repeal of the Eighteenth Amendment in 1933. Settlers also received training from the Chaffey Agricultural College, the first institution of its kind in Southern California. Toward the end of 1882, the Chaffey brothers established links with the University of California and designed a college that embodied their dreams for the state—a republic of white farmers bound together in common endeavor. The college, predicted Judge Widney, chair of the directors of the University of Southern California, would accomplish "a great mission for the race," but the institution shut its doors after two years as a consequence of lack of funds and the Chaffeys' departure for Australia.[73] The school was subsequently reorganized on a general academic basis.[74]

Ontario also represented larger trends in California agriculture. By the early twentieth century, California led the West in agricultural development. Agriculture was highly mechanized, dependent on high-value products and migrant workers, and characterized by the large farms that social critic Carey McWilliams called "factories in the field." In 1878, farmers irrigated only two hundred thousand acres of the state's farmland. By 1890, that number had grown to more than one million acres. The nation's expanding rail network, the advent of refrigerated boxcars, and Southern California's urban growth created new markets.[75]

Yet agriculture was not distributed evenly throughout the state. The transition in the 1880s from wheat farming and livestock raising to fruit production entailed a reorganization of land, labor, and capital. Fruit production required the subdivision of large estates into small, intensively cultivated plots, more labor, new marketing and production techniques, and irrigation. The decade's economic boom transformed parts of Southern California into vibrant horticultural communities, but the state's overall prosperity declined during the national depression of the 1890s. Although irrigated land had increased by more than 500 percent during the 1880s, it expanded by less than 44 percent over the ensuing ten years. Irrigation in the Central Valley declined sharply, and land monopoly persisted. California also began to lead the West in tenant farmers; almost a quarter of its cultivators did not own the land they farmed.[76]

Southern California's irrigation colonies provided a model for cooperative, corporate fruit farming and local, public ownership of water. Other piecemeal reforms such as irrigation districts accomplished less than intended but nevertheless rallied farmers to the cause of small, intensively cultivated farms. Yet these models of water development relied on local measures when centralized planning was what was really needed. By the mid-1890s, neither private nor public entrepreneurs had developed a statewide irrigation system. In the wake of the depression of 1893 and the droughts of 1898 and 1899, people turned to federal aid as the logical solution to California's water problems.[77]

### Federal Reclamation and the California Development Company

Federal reclamation of land for irrigation, municipalities, and industry symbolized the government's newfound power over western water development. Like the State Engineering Office of the 1870s and 1880s, California's federal irrigation projects of the early 1900s failed to create a statewide irrigation system. Government officials met the same obstacles that Hall and his colleagues had faced thirty years earlier: private interests that trumped the public welfare.[78] The prolonged battle between the U.S. Reclamation Service and the private California Development Company over irrigation development in the Imperial Valley along the U.S.-Mexico border revealed the government's limited power. Federal reclamation, despite its overall failure in California through the early 1920s, nonetheless served as one model for state-aided economic growth in both Australia and South Africa.

Several factors contributed to the increase in federal interest in irrigation in the 1890s. First, the 1893 depression generated the desire for an infusion of government funds into private enterprise. Tens of thousands of western farmers and businessmen lost their jobs; those already crippled by droughts in the 1880s and early 1890s suffered even more. Many Californians felt that the government had a duty to improve their standards of living and bolster economic growth.[79] Progressivism, a set of national economic, social, and political reforms that coalesced in the 1890s, also prioritized federal reclamation. At the heart of this ideology lay the belief that a highly technical society facing complex industrial problems required trained experts and centralized planning. People started to realize that not just anyone could manage the West's resources; only expert commissions could eliminate waste and inefficiency. Bureaucracies, in turn, would promote economic growth and opportunity by making available a larger share of the nation's resources.[80]

Theodore Roosevelt prioritized federal reclamation when he became president following William McKinley's assassination in 1901. Ten months later, Congress, led by Nevada senator Francis Newlands, passed the Reclamation Act of 1902. The measure provided for the construction of irrigation works on federal land, reflecting the desire for a central plan to promote small, irrigated homesteads across the country. It also created a new agency, the Reclamation Service, which reported to the U.S. Geological Survey before becoming a separate agency of the U.S. Department of the Interior in 1907. Congress authorized the Reclamation Service to conduct field surveys, erect storage works, divert water, and remove public land for irrigation. As Congress wished to help small farmers and ensure that federal water would not aid absentee speculators or large corporations, no tract could exceed 160 acres. Congress pledged 51 percent of proceeds from public land sales to reclamation.[81]

By 1907, the Reclamation Service had initiated twenty-four projects throughout the West, but because little government land in California was appropriate for irrigation projects, the state had almost none of them. The Los Angeles basin offered a good climate for fruit growing, but the San Gabriel and San Bernardino Mountains lacked adequate canyons for large reservoirs. The Sacramento Valley had a plentiful water supply and many potential reservoir sites, but most streams led into the navigable Sacramento River, which the U.S. Army Corps of Engineers governed. Joseph B. Lippincott, California's chief Reclamation Service officer, recommended only three potential irrigation projects in California: at Clear Lake; on the King's River near Fresno; and on the Colorado River, north of U.S.-Mexico border. Because the Reclamation Service feared drawn-out court battles between the government and private landowners, it completely eliminated the Clear Lake and King's River projects, leaving only one major irrigation project in California—at Yuma in the Colorado River Basin.[82]

The Colorado River runs 1,440 miles from the Rocky Mountain divide to the Gulf of California and drains portions of seven states as well as part of northern Mexico. Although small in comparison to the Mississippi and Columbia Rivers, the Colorado carries one of the world's heaviest silt loads. As it approaches the Gulf of California, the river splits into dozens of channels that broaden and block one end of the sea. Over thousands of years, the water at that end evaporated, leaving behind a crater 273 feet below sea level, the Salton Sink.[83]

The Imperial Valley is an arid but fertile region of six hundred thousand acres lying adjacent to the Salton Sink between the Colorado River and the Mexican border. The area receives an average of three inches of rain per year and can reach 120 degrees in the summer. Before white home seekers entered the region, ten thousand Cahuilla Indians supported themselves on the banks of the Colorado as fishermen, hunter-gatherers, and agriculturalists.[84] Most of the vast tracts of the rich, alluvial soil bordering the Colorado River still belonged to the federal government, which viewed this land as a prime location for a federal reclamation project. However, Lippincott and his colleagues did not foresee the conflicts that would arise with the California Development Company (CDC), which had already opened part of the Imperial Valley to irrigated settlement in 1901.[85]

Plans for the private reclamation of the Colorado Desert had begun in May 1899. Fifty-two-year-old George Chaffey, who had returned from his Australian adventure, sat down in Los Angeles with Charles Robinson Rockwood, a forty-year-old irrigation engineer, to discuss irrigation possibilities in the Imperial Valley. Soon thereafter, Chaffey, Rockwood, an Indian guide, and William E. Smythe, the era's greatest irrigation doyen, who went along to publicize the event, visited the Colorado Desert. Chaffey observed that the Colorado River's height above its surroundings enabled water to be moved westward by gravity through an ancient overflow channel called Alamo. Since the Southern Pacific's main line to New Orleans lay nearby, settlers could easily market winter produce to the East. After negotiations, Chaffey took over as the CDC's president, chief financial officer, and chief engineer.[86]

The company first needed to acquire land and water titles, which it did through loopholes in the nation's various land acts. Rockwood hired men to take up 320-acre dummy claims, which were then transferred to the CDC. Chaffey, in Ontario-like fashion, acquired water rights by founding a set of eight mutual water companies, which attached water rights to the companies' land. He transferred water stock to a holding company, the Imperial Land Company, which was run by CDC promoters. Chaffey then arranged the company's exemption from regulation by both state and federal authorities with a permit from the Mexican government that allowed him to divert water from the Colorado River a few hundred feet within Mexico, thus falling under that country's jurisdiction. Chaffey then remodeled the old Alamo Canal, which dipped into Mexico before reentering the United States, and formed a Mexican company to sell water to Chaffey's mutual water companies, which in turn resold water to the settlers.[87]

In 1902, Chaffey completed the main canal and 400 miles of lateral canals and irrigation ditches. Within eight months, more than two thousand settlers had poured into the valley. Land prices skyrocketed. The valley, with the town of Imperial at the center, soon had a hotel, bank, post office, and stores. In April 1903, the Southern Pacific built a track as far as Calexico-Mexicali, Chaffey's twin villages on the international border. The same year, settlers harvested their first crops of wheat and barley. By 1905, the Imperial Valley had an estimated population of ten thousand white settlers, seven towns, 120,000 cultivated acres, and 780 miles of canals and ditches.[88]

The Reclamation Service's interest in the valley complicated matters for the CDC and its settlers. Lippincott, who witnessed how irrigation had transformed the Imperial Valley, added it to his grand plan for reclaiming the Lower Colorado River Basin and planned to integrate the CDC's water infrastructure into a federal project. He would then construct an "All-American Canal" that siphoned water directly from the Colorado River to the Imperial Valley, bypassing Mexico to avoid international conflict.[89]

According to a Reclamation Service report, despite the challenges involved, the Lower Colorado was potentially navigable and therefore fell under federal jurisdiction. In that case, the CDC's water diversions from the American part of the Colorado were illegal. The report also provided justification for the Reclamation Service to begin designing dams, reservoirs, hydroelectric generators, and canals sufficient to irrigate ninety thousand acres downstream of Needles. Most of the project would be near or on the Yuma Indian Reservation on the California-Arizona border. The project called for the damming of the Colorado just below the Laguna Weir, which carried water into Arizona and the Imperial Valley. The Yuma reservation's lands would be opened to entry when the lands were irrigated, and displaced Indians would receive five acres each. The Reclamation Service also offered to purchase the CDC.[90]

Many Imperial Valley settlers viewed the Reclamation Service's intentions as a land and water grab. Others, unhappy with increasingly poor CDC management and water delivery, thought that the government could better serve their needs. Under Smythe's leadership, some settlers organized the Water Users' Association to raise the capital needed to enter into negotiations with the CDC for the cooperative purchase of its system. The federal Yuma project would then acquire the CDC, but this plan did not immediately materialize.

The CDC was already in disarray. Chaffey left the company after repeated quarrels with Rockwood. Anthony Heber, Chaffey's successor, rejected the Reclamation Service's offer to buy out the company. Instead, Heber secretly negotiated with the Southern Pacific Railroad for the purchase of his company. He also threatened to work with Mexican president Porfirio Díaz if the Reclamation Service intervened further. In June 1904, Heber signed a contract with the Mexican government to open a new intake on the Colorado River south of the border, which increased the water flow into the Imperial Valley. By locating the intake in Mexico, Heber also avoided Reclamation Service takeover.[91]

Building this new stretch of Mexican canal was a bad decision. The El Niño year, with its heavy rainfall, caused the new Mexican intake to quickly silt. In February 1905, the Mexican cut flooded, and a month later, more water carried away the makeshift dam. In April, the same month that Rockwood finalized his company's sale to E. H. Harriman of the Southern Pacific for two hundred thousand dollars, more floods occurred. Carl E. Grunsky, former assistant state engineer under Hall and at that time a Reclamation Service officer, examined the silted intake in June and expressed alarm — the intake had widened from fifty to six hundred feet and carried 80 percent of the river's flow into the Salton Sink.[92]

By the next year, the Salton Sink had become the Salton Sea. A pool of water eight to ten miles wide covered parts of the Imperial Valley. The Salton Sea stretched fifty miles long and fifteen miles wide. Continuing floods destroyed all of the Reclamation Service and CDC's flumes and canals and the six-foot levee around Calexico. Mexicali vanished, and the Southern Pacific lost part of its tracks. Mead, now the commissioner of the Reclamation Service, called the Salton Sea the greatest human-induced geological change in recorded history.[93]

Harriman, who now controlled the CDC, appealed to the federal government for relief. President Roosevelt, who refused to work with Mexico, held the Imperial Valley promoters, not the Reclamation Service, responsible for the disaster and declined to bail out the railroad. Roosevelt sought relief funding only after the damage had been calculated at two million dollars.[94]

The battle between the Southern Pacific and the federal government did not mark the end of the struggle to reclaim the Colorado desert. In 1911, settlers formed the Imperial Irrigation District, the largest water district in the West. The district purchased the valley's complete water-distribution network from Harriman for three million dollars. At the same time, the Reclamation Service began its Yuma project without the CDC's properties. But by 1912, the agency had irrigated only 10,500 acres. Canals quickly silted up, alkali built up, and Yuma settlers abandoned their farms. When Secretary of the Interior Franklin K. Lane investigated the project, he concluded that the failed federal undertaking revealed the government's "incapacity, wastefulness and improper administration."[95]

Lack of funding added to the Reclamation Service's problems. The agency survived on profits from the sale of land in the public domain. Yet since settlers had purchased most of California's public land by the early 1900s, the agency received few funds. Mead warned that that proceeds from western public land sales were disappearing and recommended the allocation of new funds if federal reclamation were to proceed. By 1917, the Reclamation Service had irrigated fewer than one hundred thousand acres in California.[96]

Imperial Valley farmers reluctantly joined forces with the Reclamation Service in 1919, finally permitting construction of Lippincott's All-American Canal. Yet the agency could not raise the money to build the works until the late 1930s, and the All-American Canal did not start delivering water to the Imperial Valley until 1942. The U.S. government then made a gift of land—beneath the Salton Sea—to the displaced Torres-Martinez Band of Cahuilla Indians.[97]

This gift brought this period in the history of California's irrigation to an ironic end. Between the 1870s and the 1920s, Californians reworked their assumptions about frontier development and created new economic models to fit changing conditions. Their irrigation models attempted to deal with issues such as the relative involvement of the government, private capitalists, interest groups, and the market in economic growth. Hall's plans for state-monitored private development, the Chaffey brothers' elite irrigation colonies, and federal reclamation illustrated the slow evolution and acceptance of the government as a key player in water management. Yet the varying successes of these projects showed that many Californians still

*Farm inundated by Colorado River waters, Imperial Valley, California, 1906. Courtesy of the Bancroft Library, University of California, Berkeley*

privileged local control and private enterprise over complete state control. Government bodies fared little better at stimulating agricultural growth than did private companies, which, operating in a competitive laissez-faire society, created almost insurmountable obstacles for government planners.

The overall pattern of irrigation in California through the 1920s was one of haphazard, uneven development. Engineers and policy makers nonetheless produced a valuable set of tools: expert bureaucracies; dam, canal, hydroelectric, and mining technologies; and new ideas about science, industry, rural life, property, and the relative roles of state and society in resource development. These tools did not come together as planned in California for many reasons, including the state's antiquated water laws, persistent land monopoly, and the supremacy of private over public interests. By the 1900s, despite a growing agricultural economy, Californians had yet to devise successful ways to involve the state in irrigation development and curb private capitalists in the interest of the public good.

Despite these challenges, engineers nonetheless exported their plans quickly and idealistically around the world. Californians' tools of progress— their irrigation technology, social institutions, water policy, and ideas about modern rural life—found popular support in Victoria, Australia. Victoria's landscape, frontier mentality, and other features reminded roving Californians of their home state. Yet despite the similarities, Victoria contained fundamental differences as well. These disparities posed serious obstacles for Californians' transfers of ideas about modern rural society but also provided lessons that the Californians subsequently tried to apply back home.

### CHAPTER 3

# The California Model & the
# Australian Awakening

In the late nineteenth and early twentieth centuries, California engineers experimented with different models of irrigation. Their models attempted to deal with issues such as the relative involvement of the government, private capitalists, interest groups, and the market in economic growth. Yet the overall lesson for California had been a lesson in frustration. Simply put, private interests had trumped what many people had perceived as the public good.

Despite their mixed success, Californians' irrigation experiments, technologies, and ideas about modern rural life were welcomed warmly in Victoria, Australia, which many engineers — from both California and Victoria — perceived as possessing conditions that resembled those in the American West. Alfred Deakin observed that the Australian colony was, like California, "a new country, settled by the pick of the Anglo-Saxon race, attracted by the first instance by gold discoveries, and remaining after that excitement passed away to build up a new nation under the freest institutions and most favorable conditions of life."[1] The two regions also experienced similar problems related to the growth of a new state. As in California, Victoria's recurring drought, land monopoly, and ambiguous land and water laws hindered economic growth. Settlers generally believed that Victoria's various land acts had not achieved the yeoman farmer dream that many nations shared in the nineteenth century. As international newspapers and government bulletins alerted Victorian officials to California's advances in agriculture, the Australians began to look to other regions for guidance.[2]

Although Victoria's officials knew of the Indian example, they turned primarily to California, where a common political culture was perhaps the most crucial factor in promoting transfers between the two regions. Notwithstanding the different ways that self-governing colonies and independent states operated, Victorians and Californians shared ideas about the state's role in society. The financially burdened British government forged partnerships with private companies in every Australian colony to build infrastructural works. The mix of local and state, private and public development that characterized both California's and Victoria's early years affirmed that the government should have some, albeit not supreme, influence over resource development.[3]

From the 1880s to the 1920s, a series of exchanges between California and Victoria took place, focusing both on engineering technology and on the ideas and institutions in which particular uses of engineering were embedded. Deakin, for example, visited the American West in 1885. Impressed with California's land and water policies, he returned to Victoria and adopted California's statutes. Soon thereafter, he invited the Chaffey brothers to replicate their Southern California colony of Ontario in Victoria. A decade later, under the leadership of Elwood Mead, the Victorian government implemented its closer settlement policies, a form of state-aided agricultural development that Mead subsequently tried to reproduce in California.

Despite such auspicious beginnings, neither region fully implemented the other's projects. Various factors help explain these successes and failures. Economic depression, inadequate planning, environmental conditions, and railway and market development significantly affected how these exchanges worked. These variables, in turn, helped to shape California's and Victoria's political cultures as they diverged in the early twentieth century. Victoria moved toward state socialism, while California embraced Progressivism. These different paths impaired the further cross-fertilization of technologies and ideas in the twentieth century.

### The Australian Awakening:
### The Royal Commission Visits the American West

Severe droughts plagued southern and eastern Australia, including Victoria, between 1881 and 1886. The stress caused by such water shortages raised public awareness of the need for water conservation and irrigation. During that time, Deakin, then Victoria's minister of public works and water supply, began to study the possibilities of irrigation, and California emerged as the most compelling model.[4] "*Ergo*," a local newspaper asked, "if the Californian is [irrigating] successfully, why should not the Victorian? . . . Here is a bread-basket ready to supply the hungry millions of the old lands with corn, wine and meat."[5]

In 1883, Deakin, drawing heavily on the California experience, introduced the colony's first irrigation legislation. The Water Conservation Act of 1883 created irrigation districts like those then pending before the California State Legislature—a decentralized system of water users monitored by the state. Deakin placed California's bill almost directly into Victoria's statutes. Thus began the great "Australian awakening," Victoria's foray into irrigation agriculture.[6]

Agricultural development had of course evolved since the first English convicts landed at Botany Bay in 1788. Their experiences in Liverpool and London did not prepare them for Australia's sandy scrubland, thin forests, and such strange natives, plants, and animals as bushland Aboriginals, eucalyptus, wallabies, and platypuses. Yet English colonization persisted, and the British Crown carved Australia into territories in the 1830s. Two decades later, discoveries of gold in Victoria, Western Australia, and New South Wales attracted immigrants from all over the world. Disease and displacement had reduced the aboriginal inhabitants by two-thirds by the end of the decade. English settlement across the Pacific was a fait accompli.[7]

The colonial governments of Victoria, New South Wales, and South Australia invested in railroad, banking, and land development from the start, as the colonies' small populations lacked capital for their own projects. The colonial states also remained landlords over most of the territories, even though most of the best land passed into private hands.[8]

The region's semiarid climate did not bode well for agricultural settlement. Still, British immigration to Australia continued throughout the nineteenth century, and colonial legislators encouraged the entry of newcomers into agriculture. A series of Land Selection Acts passed between 1860 and 1880 proposed to settle small, capitalist farmers on 320 acres of land each. Many of these acts cited the U.S. Homestead Act (1862) as a model.[9] Would-be cultivators selected more than 1,000,000 acres in Victoria alone, but the legislation did not produce an even distribution of land. Some settlers abused the loose land laws. Others lacked access to transportation and badly needed water storage facilities. Agricultural settlement proceeded slowly.[10]

More than half of Australia receives less than fifteen inches of rain per year, too little for perennial agriculture. Victoria, New South Wales, and South Australia share the Murray River, the region's only regularly snow-fed river. Stuart Murray, chief engineer of the Victorian Water Supply Commission, observed on a trip down the river in 1898, "With water, everything is possible in this climate, in the way of fruit culture or agriculture. Without this everything is doubtful."[11] During the 1880s, colonial officials began to correct what one man called this "great defect in nature."[12]

Deakin, Australia's "father of irrigation," was one of the first politicians to understand the implications of this "defect." Born in Melbourne in 1856, Deakin rose to become one of Australia's most distinguished statesmen. He was admitted to the bar in 1878 and entered politics shortly thereafter as a member of the Liberal Party. At age twenty-six, he became Victoria's minister of water supply. He subsequently worked toward Australian Federation, which took place in 1901. He served as Australian prime minister three times in the first decade of the commonwealth and produced much of its founding legislation, including the Immigration Act, or white Australia policy.[13]

Although Deakin is now remembered primarily for his years as prime minister, he addressed important resource issues in his earlier political career. In the 1880s and 1890s, he pioneered much of the legislation that provided for Victoria's irrigation and land policies, implementing those that helped transform parts of the arid region into a productive garden landscape.[14]

As the mining industry started to deforest large areas, silt rivers, and impede agrarian settlement, Deakin and other colonial administrators began to look at other countries' models of resource development. By the 1880s, Australian engineers had thoroughly studied America's and British India's irrigation works. India's Public Works Department provided the best technical model for state-administered water development, while California served as an example of local control with minimal state aid. "We are fortunate," Deakin wrote, "in lying between these two greatest fields of modern irrigation, and should reap many advantages from an intimate relation with both." For the most part, California's model prevailed. California's geographical and political similarities to Victoria made the U.S. state more relevant; more important, Victorians believed in private enterprise guided by public institutions rather than in imperial control, as in India.[15]

In the early 1880s, the Victorian government began to send emissaries to California to learn more about irrigation. Hugh McColl, a member of the Victorian House of Parliament, visited California in 1884. After meeting with state engineer William Hammond Hall, McColl concluded that California had made the most rapid strides in the matter of irrigation.[16] In 1884, Deakin convinced the Victorian cabinet to appoint a royal commission to visit the American West to study irrigation and agriculture. In the spring of 1885, Deakin sailed to America, accompanied by John L. Dow, a member of Parliament (and later Victoria's minister of agriculture), and Sir Edward Cunningham, chief reporter for the *Melbourne Argus*.[17]

The Victorian delegation spent thirteen weeks visiting American sights and meeting with as many engineers and scientists as possible, including Hall and Wyoming's state engineer, Elwood Mead. Traveling by rail, the visitors gained insight into America's "triumphal march of progress," celebrating the "homesteads and harvests and cities of men" that rose like phoenixes from the land. If the Americans could succeed in building a civilization where nature had failed, the Australians believed they also could do so.[18]

Deakin's party arrived in New York in early March 1885. They were astonished by Brooklyn's suspension bridge, Manhattan's cable cars, and the elevated railroads that ran along Broadway. From New York, the three men made their way to Boston and Concord. A week later they entered Chicago, the city of great skyscrapers. They then traveled through the Rocky Mountains to Greeley, Colorado, before reaching Salt Lake City, where they examined the town's irrigation colonies. The Mormon Church exercised a powerful influence in producing military-style cooperative endeavors. At the end of March, the royal commission crossed the Sierra Nevada into California.[19]

Deakin concluded that Victoria and the American West, with their similar climates, soils, and rainfall, were "as complete as such parallels can be."[20] As he traveled around the state, he gathered a wealth of information. He noted the conflicts generated by different water rights laws and the varying responsibilities of levels of government, local communities, and individual irrigators. He collected data on soil, types of crops, precipitation, and water sources.[21] Victoria's lands were more fertile, but the Sierra Nevada's streams were more numerous and adaptable. Whereas the Sierra Nevada's heavy snowpack fed the California system, Australia's scant rivers, including the Murray, dried to a trickle in the summer. California's shallow rivers required only light structures to raise the water. Engineers had more difficulty raising and diverting the deeper rivers of northern Victoria. As a whole, California claimed more easily engineered rivers.[22]

The Royal Commission also visited farming settlements in Southern California. At the time, Los Angeles was an adobe town with a population of about twelve thousand and a few Spanish missions. After meeting with George Chaffey, the royal commission deemed Ontario a "model colony," a name that stuck. Cunningham and Dow convinced Deakin that they would be fortunate to start a similar settlement in Victoria.[23]

In early April, the Australians left Los Angeles and traveled to San Francisco. There, Deakin met with Hall, the "most competent judge of [irrigation] questions in the West."[24] Deakin and Hall spent a few days examining the vineyards and irrigation colonies around Fresno and Merced, and the two discussed Victoria's irrigation possibilities at length.[25] Impressed, Deakin invited Hall to Victoria to design a comprehensive irrigation regime there. Hall, embroiled in political issues, declined but agreed to consult for the Victorian government, a relationship that continued over the next few decades.[26]

After returning to Victoria, Deakin published a report on his travels, *Irrigation in Western America*. Overall, the report criticized the American West, despite its seemingly content settlers and abundant orchards. Deakin noted that California's state government invested almost nothing in the long-term commercial success of its farmers, a policy that led to speculative waterworks. Nor had the state reconciled diverse group interests. He observed that developers built canals without sufficient capital and without regard to soil type, the possibility of unprofitable crops, and market reliability. "The crippled condition of irrigation enterprise," Deakin wrote, "remains an impressive warning as to the necessity for providing against those complications in a new State before vested interests become too deeply involved." In fact, Deakin believed that none of California's legislation — except for Hall's irrigation districts — had any value for Victoria. Above all, Deakin wanted to avoid the conflicts over water rights that had divided Californians. Yet he remained optimistic about California's future.[27]

After visiting California, Deakin proposed a more central role for the government in Victorian irrigation. He believed that state action was often necessary and sometimes unavoidable, particularly in a sparsely populated, capital-poor region. Deakin predicted that the infusion of public funds for large-scale irrigation works would eventually lead to a rise in agricultural prosperity. More pressingly, Victorian politicians desperately needed to do something to alleviate the drought.[28]

Deakin's faith in state aid drew on different strands of the land reform philosophy that was popular at the time, including that of California social critic Henry George, who challenged the California dream of democratic, landowning farmers. California's beauty, George believed, defied the corrupt foundations of society. His *Progress and Poverty* (1879), one of the best-selling economic treatises of all time, proposed a solution to California's land monopoly, low wages, and poverty: a single tax on land, which would help to redistribute the state's wealth. The single tax movement also promoted utopian ideas and moral ideals such as temperance. Although never fully adopted, George's ideas elicited popular support around the world.

In 1890, George toured Australia, where he was warmly received by politicians, professors, and land reform groups.[29]

At least in theory, Deakin supported George's single tax idea. Deakin sought to discourage private speculation, subdivide large properties, and increase land values for the community after irrigation started. This interest in land reform could have translated into socialist policies, but the Australian middle class embraced George's theories. Nonetheless, his ideas failed to take root in both California and Australia because they did not garner the support of the labor movement. Still, George's philosophy, Hall's more practical recommendations, and other foreign influences strongly influenced Deakin's Irrigation Act of 1886.[30]

This measure, which passed by a large majority, was a hybrid that favored a mix of local and state control. Its most radical aspect came from India. Deakin recommended that the state exercise supreme control, even ownership, of all rivers, lakes, streams, and other sources of water. He drew not only from the American West but also from the Spanish Law of Waters (1866) and the Northern India Canal and Drainage Act (1873). The 1886 act marked a new era in the history of water legislation. It vested all water rights in the Crown, prevented the acquisition of further riparian rights, authorized the government to construct national irrigation works, and provided for the creation of local irrigation trusts.[31]

The act also reflected Deakin's confidence in individual enterprise: he wanted local landowners to initiate schemes with some form of state guidance, just as they did in the American West. The act's irrigation trusts drew on the California model of local control. Engineers designed the trusts along contours of the river basin and enabled water users to carry out their plans with money advanced by the government. The 1886 act also promoted various private schemes under the Waterworks Construction Act (1886). The most famous of these private ventures were the Chaffey brothers' irrigation colonies, Mildura and Renmark.[32]

The ninety irrigation trusts established in northern Victoria formed the heart of Deakin's decentralized state-aid program. In 1887, the government allocated its first funds toward the construction of the Goulburn Weir along the Murray River, which supplied the Rodney Irrigation Trust. In the next decade, boosters projected that the Goulburn Valley would be the new "Eldorado of agricultural wealth."[33]

Deakin's attempt to place the technical and political responsibilities for water supply with local bodies backfired. About 80 percent of the trusts, particularly in northern Victoria, suffered from inadequate water conservation. Seasonal uncertainty and haphazard regulation provided less water than predicted.[34] Financial mismanagement also crippled the trusts, as many settlers defaulted on their loan repayments to the government. Few of the trusts even met operating costs. An 1896 royal commission sent to evaluate these problems concluded that many of the trusts could not be justified financially. In 1899, the Victoria legislature passed a relief act that wrote off three-quarters of the trusts' existing liability.[35]

Although Deakin favored government aid, he also wanted to try out the private colonization enterprises he had witnessed in the American West. The 1886 legislation authorized Victoria's most ambitious private irrigation scheme to date: an irrigation colony started by the Chaffey brothers. Despite some opposition to what some Victorians viewed as the "Yankee grab," Deakin justified the Californians' private land acquisitions in Victoria on the grounds that the colonies would solicit international capital and expertise. Victoria thus began its first great experiment in private land colonization.[36]

### Won from the Wilderness: Australia's Irrigation Colonies

In the spring of 1886, thirty-eight-year-old George Chaffey and his thirty-year-old brother, Ben, traveled to Australia. As part of its new irrigation policy, the Deakin government granted the brothers land along the lower Murray River. The Chaffeys established their settlement at the old, run-down cattle station of Mildura and reproduced Ontario, their Southern California colony. As much as they were engineers, the Chaffeys were also social planners whose ideas in some sense preceded those of the American Progressive movement in the first decades of the twentieth century. The brothers, like their Progressive colleagues, pinned their hopes for a simpler, individualistic social order on technical change.[37]

The contract that George Chaffey signed with Deakin set aside land for development twelve miles from the Victoria–New South Wales border. Deakin granted the brothers 250,000 acres under a twenty-year license. They could sell land for orchards in blocks of fewer than 80 acres and for other agricultural purposes in blocks of 160 or fewer acres. The government also granted to the Chaffeys a large share of the waters of the Murray River to ensure that at least a half million people would eventually settle there.[38]

The Chaffeys' land stretched along the Murray River in Victoria's northwest corner, the heart of the barren Mallee country. The region comprises more than eleven million acres of land, and its semiarid climate contrasts sharply with the rest of Victoria's milder climate and forested landscape.

Despite the wet month of December, many parts of the Mallee region experience chronic drought. One traveler described the landscape as "a hissing Sahara of hot winds and red driving sand, a howling, carrion-polluted wilderness" of drought, kangaroos, and emus.[39]

The Murray River originates in the southeastern Snowy Mountains and receives the snows that fall on the mountains' seven-thousand-foot peaks. Like California's Sierra Nevada rivers, the Murray, which flows thirteen hundred miles through South Australia and divides Victoria and New South Wales, does not depend entirely on rainfall. It drains an area of approximately three hundred thousand square miles—five-sixths of New South Wales, more than half of Victoria, and one-seventh of the entire continent. The river feeds Mildura's plain of rich, red, sandy loam.[40]

In the fall of 1887, the brothers set about replicating Ontario. They formed the Chaffey Brothers Limited Company and incorporated the Mildura Irrigation Company, a mutual water company modeled after Ontario's. They planted fruit and gum trees, the same species that lined Ontario's Euclid Avenue, around wide boulevards, subdivided Mildura with Monterey cypress and olive hedges, and built townhomes and villas. Horticultural lands abutting the irrigation channel stretched in orderly plots beyond the homesites. Redbrick cottages, garden-enclosed offices, and green lawns greeted colonists by 1890. The Chaffeys also built two banks, a progress association, a hotel, a church, a telegraph and postal office, and a school.[41]

The colony lacked only a nearby railway, which the government had promised to build. Mildura lay more than two hundred miles away from the nearest railheads at Warracknabeal and Wycheproof. In America, private corporations constructed railways, and settlers followed. In Australia, the government waited until settlers showed that they could keep the railway, a public venture, alive. Paddle steamers transported local freight up and down the Murray.[42]

Through clever advertising campaigns, the Chaffeys received applications for land purchases before the colony existed. The "Red Book," a crimson, gilt-lettered, and richly illustrated volume published in London by J. E. Matthew Vincent in early 1889, helped draw English settlers to Mildura by advertising the Mallee as the new "promised land."[43] Enthusiastic accounts of life in Mildura, together with maps of the town, orchards, and proposed railway route, enticed would-be colonists who sought a simpler, more natural lifestyle. Romantic poets, including Ruskin, Tennyson, and Burns, waxed about a sublime nature that would uplift the hard lives of urban workers. They called for the teeming multitudes of the Old World to people Australia's vast solitude. Vincent included glossy pictures of Mildura at its most inviting—sylvan groves, the misty Mallee, the willowy river. Neat cottage homes and dense orange groves completed the picture.[44]

The Red Book lured a few thousand people away from London, Manchester, and Liverpool. Many future settlers viewed Australia the same way that easterners in the United States had viewed the American West in the 1850s and 1860s. Because of its great expanse of seemingly free land, the West had given rise to the hope of new beginnings for eastern urbanites and farmers. The Chaffeys advertised Mildura, too, as a safety valve for the overflowing laboring classes, a place where democracy and social mobility could flourish.[45]

The Chaffeys portrayed Australia as the panacea for England's human suffering: social dislocation caused by enclosure acts passed a century earlier, unsuccessful "right to live legislation," and the population growth and food shortages discussed by Malthus, Marx, Mill, and Darwin. But across the ocean, land colonization, the application of technology to further social goals, and the expansion of the English frontier would fulfill the "progressive and enterprising purposes" of the Anglo-Saxon race.[46] The Chaffeys assumed that the subdivision of large estates into smaller ones, generous credit systems, and close-knit community life would transform the wage worker into the romanticized farmer of yore. Rather than engaging in "cynical and selfish exploitation," colonists would partake in "enlightened, honourable, energetic, equitable business."[47] Francis Myers of the *Melbourne Argus* likewise believed that scientific irrigation would alleviate city dwellers' "debauched and degraded" state.[48]

Like the great company town directors of Lancashire, Manchester, and Pullman, the Chaffeys readily bestowed their paternal—and commercial—guidance on their colonists. Mildura, both a cooperative and commercial enterprise, catered to settlers who possessed a strong capitalist ethos. The Reverend A. R. E. Burton, the Chaffeys' spokesman, argued that with hard work, settlers could extract gold from the Mallee's fertile soil.[49]

During the 1880s land boom, the Chaffeys sold irrigated blocks of land to colonists from around the world. Just as its founders anticipated, Mildura attracted settlers from England's overcrowded cities, even if the new arrivals generally did not represent the mixed socioeconomic group the proprietors had expected. Most immigrants came from the professional classes that could afford to start a new life halfway across the world or were English gentlemen who wanted to dabble in rural life. Only a few colonists came from poorer parentage. Nevertheless, in the early 1890s, Indian army majors, mill managers from Manchester, bank managers from London,

*Mildura settlers. Courtesy of the State Library of Victoria, Australia*

deep-sea captains, accountants, engineers, and farmers settled into rural life. Former bankers and longshoremen rubbed elbows, creating a camaraderie that belied the conflicts between labor and capital sweeping through Europe and America. By 1892, Mildura boasted more than four thousand people.[50]

Social mobility depended not only on landownership but also on settlers' ability to transform the barren Mallee into a productive garden. "Nature," the Chaffey brothers concluded, needed only to be "encouraged — educated, in fact."[51] Technology schooled nature. After the Chaffeys laid out the town, they began to clear the land. They imported the world's most modern machinery: powerful steam plow engines and traction engines, the Mallee roller and stump-jumping plow. These machines enabled settlers to clear ten-acre blocks in one day.[52] One 1888 visitor found his soul stirred by "watch[ing] the busy engines . . . one tearing up mallee bushes or pulling out big gum stumps, another hauling the merciless scarifier with its nine great spikes, through . . . mile after another of primeval blue bush . . . till a fair, even, reddish tilth of sandy loam takes the place of the tussocky wilderness of a few hours before."[53] The substitution of steam power for bullocks, horses, and manual labor, Myers claimed, severed humankind's "barbarous past" from its "civilised future."[54]

Mildura presented the Chaffeys with technical problems that had not been present in California. The brothers had constructed gravity irrigation systems to carry water down from the Sierra Madre Mountains to Ontario.

In Mildura, the Murray River ran far below its banks, meaning that water had to be pumped up to the surrounding countryside. The Chaffeys overcame this difficulty by designing pumps that ran on hydroelectric power, the first of their kind in Australia. By 1892, eleven pumping plants and more than 450 miles of canals, siphons, flumes, and bridges crisscrossed Mildura.[55]

By then, the Chaffeys had sold seventeen thousand acres of land, sixty-five hundred of which had been planted with grapes, apricots, oranges, lemons, peaches, and olives.[56] "Talk about the romance of engineering!" proclaimed an American editor, who went on to compare the Chaffeys to Moses, striking the rock to produce water.[57] Mildura held the same promise for colonists as Canaan, the biblical land of milk and honey, had held for the Jews during their forty years' trial in the desert. By the 1890s, Mildura's garden homes, orchards, and cooperatives suggested that civilization, as one journalist commented, had been "won from the wilderness."[58]

Social engineering accompanied this great transformation of landscape; the Chaffeys hoped that their application of technology to nature would also produce "an example and blessing to the civilised world."[59] The first experiment in social planning involved the prohibition of the sale of liquor. The colony founders believed that temperance would attract a good class of Christian settlers who would become a reliable, devout labor force.[60] But despite its lofty ideals, temperance agreed with few colonists. Local public opinion made it impossible to enforce Prohibition in Mildura, and underground alcohol sales persisted. By 1893, settlers agreed that while their colony should adhere to the ideals behind temperance, the Mildura Settlers' Club and other premises could sell alcohol.[61]

The Chaffeys also focused on scientific education, which would lead to better farming methods and hence greater profits and happiness. As in Ontario, the Chaffeys set aside land in Mildura for an agricultural college, but no facilities had been built by the mid-1890s. The settlers exacerbated the situation: many of the "gentlemen" who had previously romanticized the benefits of rural life resisted the Chaffeys' attempts at education.[62]

Despite these glitches in the production of a model rural society, Mildura flourished for two years. But in 1891, the colony succumbed to national economic depression. Victoria's decadelong land boom collapsed, precipitating a financial panic that turned into the worst depression in Australian history and culminated in the great Bank Smash of 1893. As depression set in, public works, including railways, came to a halt. Mildura fought off bankruptcy for two years. Chaffey Brothers Limited survived, although the crash deprived Mildura of its private capital and destroyed its local market for fruit. By 1895, the colony was almost in receivership.[63]

*Construction of siphon pipelines and channels. Courtesy of the State Library of Victoria, Australia*

The depression only heightened tensions that already existed within the colony. The Chaffeys had overextended themselves and Mildura from the start. Mildura occupied many times the area of Ontario and posed complicated technical problems that Ontario lacked. George Chaffey also developed the colonies of Renmark in South Australia, the Weribee project near Melbourne, the Mulgoa scheme in New South Wales, and the beginnings of another irrigation colony in Queensland. With funding short, these projects proved too great a burden for one man.[64]

Local complaints added to these problems. In 1895, a group of farmers charged Chaffey with mismanagement. They had failed to pay interest on loans obtained from the government for the cost of operating their waterworks and blamed the Chaffeys.[65] In May 1896, the Victorian government appointed a royal commission, composed mostly of Mildura's foes, to investigate the situation. Many of the commission's findings ran counter to the enthusiastic testimony provided by a minority of the local farmers, prominent politicians, and expert engineers, including Murray and Deakin, Mildura's loyal supporters. Their expert opinions did not combat the commission's damaging final report.[66]

Many colonists complained about their plots of land, their poor agricultural education, the inadequate water supply, and the Chaffeys' Yankee greed.[67] James Henshilwood, who had no horticultural experience, settled on ten acres in 1889. Within three years, salt and seepage destroyed his land.[68] John James Thompson Lever, lured from England to Mildura by the Red Book, bought a block of land sight unseen, planted orange trees on bad soil, and lost almost all of his initial investment. Yet unfortunate circumstances did not affect everyone. William P. Buckhurst, a Californian who had settled in Mildura, saw only "a prosperous community."[69]

Even in the best of circumstances, Ontario's success would have been difficult to replicate. Settlers' accusations revealed just how greatly Australian national ideology had started to diverge from the historic American pattern of minimal state development, though the United States changed as well. In the 1890s, American politicians began to implement the Progressive reforms that characterized early-twentieth-century bureaucratic centralization. Victoria's political insecurity during the 1880s also reflected the colony's changing stance toward the state. Some political parties, including Deakin's Liberals, began to move away from laissez-faire toward government aid and piecemeal government ownership.[70]

Poor transportation and communication networks also contributed to Mildura's failure. More than a thousand miles of railway crisscrossed Victoria in 1880, but Mildura had no rail line that would open up the Mallee for settlement and trade. The government lacked the funds to build a line, and, more important, most politicians did not wish fully to support private enterprise. Colonists thus relied on the Murray River, whose seasonal nature posed many obstacles for trade. The Chaffeys held a controlling interest in the river's paddle steamer service, but Mildura faced the risk of being totally cut off from water transportation in drought years. A Mildura maker of jam testified before the royal commission that the river's summer closing made it "impossible to sell . . . fruit of any kind."[71]

The Chaffeys also failed to gauge Mildura's wine and fruit markets, optimistically assuming that Victoria, like California, possessed an insatiable domestic market.[72] The Royal Commission concluded otherwise.[73] Although two-thirds of the population resided in Australia's most fertile lands in Victoria and New South Wales, the continent had barely three million people in 1890.[74] Mildura settlers glutted their own markets. One inhabitant who owned twenty fruit-producing acres had three tons of unsold fruit in 1895 alone.[75] The nearest foreign markets lay half a world away, and worldwide depression made fruit consumption a luxury.

In the mid–nineteenth century, Australians exported the bulk of their agricultural products, including their chief export, wool, to Britain. A few decades later, America, Russia, Argentina, and India pushed Australian wheat out of the world market. Once refrigeration became widely available, Melbourne supplied the London market with pears and apples during the summer months. But between 1885 and 1890, agricultural products, including butter, fruit, wheat, and sugar, still comprised less than .5 percent of Australia's total exports. Through the 1890s, gold remained Australia's most remunerative industry.[76]

Despite some colonists' continued support, the royal commission's Mildura report ruined George Chaffey's career in Australia. Bankruptcy followed, and he returned to California in June 1897. The commission's verdict, however, did not withstand the test of time. The return of economic prosperity to Australia brought renewed life to Mildura. Deakin believed that the Chaffeys had merely entered during a time of economic and political uncertainty. In the absence of such "untoward circumstances," Deakin wrote to Ben Chaffey, Mildura might have succeeded.[77]

Despite these shortcomings, general praise for the Chaffeys followed in the royal commission's wake. As Murray summarized, the Chaffeys' "work has done more than any other simple enterprise to elucidate the problems that surround the great generation of the settlement of the Australian Continent by an industrial European population."[78] That said, Mildura's technical innovations outlived much of the colony's initial social context: temperance, agricultural education, and cooperative farming endeavors faded into the past.

But other facets of the colony flourished. Although George Chaffey left Australia, Ben and his family remained in Mildura. The Victorian government delivered its promised railroad in 1903, attracting more settlers, and started to subsidize the colony. In 1911, Ben incorporated the successful Mildura Winery. The dried vine fruits industries also prospered. The Chaffeys had not only reclaimed the landscape but ultimately engineered a profitable community that showed how progress occurred despite depression and squabbles over land, water, and the role of private capital and government.[79]

### Elwood Mead and Closer Settlement

The 1890s and 1900s heralded new ideas about the role of government in irrigation development. A view emerged that the state government should actively aid private enterprise. The growth of closer settlement, a set of policies designed to settle barren lands in Victoria, showed how state funds could secure great agricultural profits. Mead, who headed Victoria's State Rivers and Water Supply Commission between 1907 and 1914, instilled the closer settlement program with the principles of the American Progressive movement: sound business administration, scientific management, and more than a little idealism about modern rural life. The success of closer settlement, though limited to certain regions, suggested that government projects could significantly increase agricultural growth.

Closer settlement had existed in Victoria since the 1860s, when the government unsuccessfully tried to condense wheat-growing lands. The movement elaborated on the idea of designing farm units of appropriate size to enable families to live comfortably in a community. Although usually more powerful in rhetoric than in reality, closer settlement in Australia and New Zealand between the end of the nineteenth century and the start of World War I seemed to generate the conditions for upward mobility. Murray's 1904 Closer Settlement Act gave the Victorian government the power to acquire and subdivide large estates.[80]

The growth of closer settlement evolved as part of a much wider process of governmental influence. Several factors contributed to the government's interest in irrigation development. First, the economic depression of the early 1890s hindered private enterprise, as Deakin's failed irrigation trusts indicated. Second, the spirit of social reform that characterized Federation also encouraged land reform. New government agencies, such as the State

Rivers and Water Supply Commission, assumed responsibility for irrigation development. Federal officials now worked with new state agencies to divide the use of the Murray River among three Australian states. Overall, Federation shifted water policy away from Deakin's decentralized system to a program of state control and ownership.[81]

Though known as "state socialism" for irrigated settlement, closer settlement was not related to international socialist movements. Instead, like the American Progressive movement, it reflected a new interventionist role for government. The state acted as public entrepreneur, expert, and consultant to the private sector. In Mead's view, the government that inaugurated such development "is nearer a real democracy than a Government that leaves the settler to struggle unaided."[82]

The American West again served as a starting point for the Australian government's new role in rural life.[83] The growth of resource agencies in the 1890s preceded Federation by about a decade, and Australia drew on many of the U.S. examples. The Reclamation Act of 1902, a symbol of the new U.S. power in resource development, may have served as a model for the Victorian Water Act of 1905, which created the State Rivers and Water Supply Commission.[84] The act, introduced by the new minister for water supply, George Swinburne, authorized the agency to carry out surveys; plan, construct, operate, and control water supply works; and implement reclamation works for flood protection, irrigation, and drainage. The act also restricted riparian rights to domestic purposes and stock raising. The beds and banks of all streams became property of the Crown, as was the case in British India, thus completing the nationalization of sources of water supply that Deakin's Irrigation Act of 1886 had started. The Water Act also replaced the decentralized irrigation trusts with state-run irrigation districts. Only Mildura, which had been created under different legislation, remained.[85]

In March 1907, Mead, one of California's preeminent engineers and proponents of planned irrigation, replaced the aging Murray as chair of the State Rivers and Water Supply Commission. Mead, born on an Indiana farm in 1858, had completed his education in science and engineering at Purdue and Iowa State University. He then worked for the U.S. Corps of Engineers before accepting a teaching position at Colorado State Agricultural College, where he became the country's first professor of irrigation engineering. In 1888, he left Colorado to become Wyoming's territorial engineer. When the region became a state two years later, Mead drafted the part of the state constitution dealing with water rights, ending legal conflicts by vesting water rights in the state. Mead served as Wyoming's state engineer until 1898. Although many western states considered Wyoming's water laws to be model statutes, few states adopted them. By 1897, Mead started to believe that federal action was necessary to realize the full benefits of reclamation and began to look for new opportunities.[86]

In 1899, Mead was appointed chief of the Office of Irrigation Investigations, a new branch of the U.S. Department of Agriculture. Mead's appointment put him on the national scene at a time when the government was starting to embrace the idea of federal reclamation. During his eight years with the Agriculture Department, Mead also served as professor of rural institutions and irrigation at the University of California at Berkeley. He founded the Department of Irrigation in 1901, published the standard textbook on irrigation in 1903, and received an honorary doctorate of engineering from Purdue in 1904.[87]

Mead held contradictory views on private, state, and federal reclamation, as his work in California and Victoria revealed. He feared that making irrigation a federal responsibility would threaten individual states' control of water. At the same time, he believed that Washington had an obligation to lead the settlement of the rural West and irrigate public lands. At the Department of Agriculture, Mead disagreed with prominent policy makers over plans for a federal water development program, instead favoring states' rights and federal irrigation only of public lands. His position met with opposition from powerful supporters of federal reclamation such as Frederick Haynes Newell and George Maxwell, and President Theodore Roosevelt ultimately rejected Mead's stance.[88]

During the bitter and prolonged controversy over federal reclamation, Mead received an invitation from the Victorian government to assume the post of chair of its new State Rivers and Water Supply Commission. Arriving in Melbourne in March 1907 with his wife and five children, he soon began to move Victoria away from local, private irrigation development to state-aided agricultural settlement.[89]

Although Mead strongly disagreed with the principle of federal sponsorship of irrigation schemes, he acknowledged that only the government, preferably at the state level, had the means to encourage irrigation, immigration, and intensive land settlement. He viewed state aid as an initial, necessary phase to create model settlements and restore the small family farm, the cornerstone of a rural democracy.[90]

Mead thus diverged from Deakin's nineteenth-century liberalism. Focusing on business management, Mead radicalized the principles behind previous articulations of closer settlement. Irrigation works did not automatically create irrigated agriculture, of course. Money spent on dams and canals led to even greater expenditures for houses, farm buildings, and fences before irrigation works paid off. Mead argued that the state should purchase irrigable properties in every district, divide them into smaller, family-size farms, and assist immigrants until they became established farmers. Mead's closer settlement program centered on the radical idea of rural credits. A state-run and -financed banking system would provide loans that would enable home seekers to buy and improve farms. Settlers would pay off their debts in thirty-one years, with interest rates set at half the rate paid by farmers in the American West.[91]

The American experience provided valuable guidance for the closer settlement program. Between 1905 and 1910, Victorian dignitaries and journalists visited California's irrigated settlements, concluding that most created cohesive communities and excellent infrastructure: roads, rural mail delivery, newspapers, trolley lines, even railroads. "All recent American experience," one reporter noted, "supports the decision of the Victorian Government to divide the best of the irrigated areas under its control into small allotments."[92]

But Victoria could improve on California's colonies. An American farmer required a lot of capital to start and maintain a family-sized farm of fifty acres. Victorian state aid enabled men of lesser means to do so. The government, for example, paid for four-fifths of immigrants' steamship passage to Melbourne. Once in port, Victorian officials greeted settlers and helped them acquire loans from the government and purchase the already-cleared irrigation blocks. Colonists then settled in ready-made communities with irrigation works, roads, schools, churches, close townships, transportation, and markets. And as in the Chaffeys' irrigation colonies, the rhetoric of modern rural life flourished. "Old methods must perish," the *Victorian Settlers' Guide* noted, making room for "a more scientific system" of landholding governed by the state, "a tender-hearted landlord to the *bonâ fide* settler."[93]

During his tenure as chair of the State Rivers and Water Supply Commission, Mead supervised thirty-two closer settlement projects. His first selections of land included the northwestern Mallee areas of Merbein, Swan Hill, and Red Cliffs (adjacent to the First Mildura Trust). Victoria's Greater Western District scheme alone proposed the purchase of eight estates occupying 1.3 million acres to supply the needs of five thousand families. Farms varied in size from 18 to 100 acres, depending on the type of crop.[94]

In spring 1910, the government launched its largest advertising campaign to date, appointing Mead and Hugh McKenzie, Victorian minister of lands, to the Victorian Land Settlement Delegation. The group spent six months touring Italy, the United Kingdom, Denmark, Holland, and the United States, inspecting irrigation colonies and small farms and advertising the great possibilities of Victoria's closer settlement schemes through public addresses. Although the delegation's colonizing efforts abroad came to nothing in Italy and America, they achieved more success in Britain.[95]

A strong economy, with rising prices for agricultural products, accompanied the expansion of closer settlement in northern Victoria after 1909.[96] More than six thousand people, mostly English immigrants, had arrived seeking land, and one-tenth of Australia's wheat belt (2.2 million acres) lay within the closer settlement area.[97] In 1911, Mead boasted that one district with 390 families had shown a twentyfold increase in population in two years. Eighty-one homesteads occupied a 38,000-acre estate in another region.[98] One of the settlement inspectors, formerly a successful farmer in California's fertile Imperial Valley, claimed that Victoria had made more progress in eighteen months than California's Imperial Valley had achieved in five years.[99]

Yet many of the closer settlement initiatives did not profit during Mead's tenure. Nor did most succeed in the long run. The *Melbourne Age* claimed in 1913 that Victoria lost about eleven hundred pounds a week from its irrigation expenditures, and farmers struggled to find markets.[100] In 1915, Mead argued that the Crown had extended its program far enough to provide the necessary incentive for private development.[101] A minor 1920s recession and major 1930s depression threatened irrigated and nonirrigated farms alike, and other Australian states competed by offering free land. But since the government had established the foundation for intensive agriculture in the 1900s, economic conditions did not destroy the closer settlements as they had the earlier irrigation trusts. After World War II, closer settlement became a way to place returned soldiers on the land.[102]

Mead started to think about leaving Australia in 1913. Although he felt that he had not fully reached his goals there, he believed that he had blazed a trail that others could follow. A year later, Mead received an offer from U.S. Secretary of the Interior Franklin K. Lane, who was seeking a replacement for the controversial Newell as director of the Reclamation Service. Mead turned down the position but left Australia anyway.[103]

In 1915, Mead joined the Department of Rural Institutions at the University of California at Berkeley, where he received strong support from those interested in improving rural life in California, including university president Benjamin Ide Wheeler, California senator John D. Works, and the state's governor, Hiram V. Johnson. Though Mead's interests spread over a great range of agrarian issues, he most hoped to apply the Australian plan of state-aided settlement to California. The great lesson for the United States, he believed, was that the Australians had acted "wisely and efficiently in carrying out great works for the common good," though their land settlements were far from successful in the long run.[104] Nonetheless, Mead believed that adoption of the Australian models would help to reverse the decline in the quality of American rural life.[105]

## Elwood Mead's Demonstration in Scientific Colonization

In 1915, the exchanges between California and Victoria that had started with Deakin's visit to the American West came full circle with Mead's plan to create state-aided irrigation settlements in California. If not enthusiastically embraced by large landowners or private entrepreneurs, Mead's statist land and water proposals gained general acceptance from the public. Under presidents Theodore Roosevelt, William Howard Taft, and Woodrow Wilson, Americans witnessed radical changes in the way that government operated in relation to society. Progressivism introduced a new bureaucratic system, a way of reordering government to accord with the modern home and workplace. A loose collection of reforms, the Progressive movement produced policy that ranged from Prohibition and antitrust legislation to new labor laws.

Natural resource policy and conservation played an integral role in the process of centralization and expert planning at the state and national levels. Both evolved as institutional responses to the rapid pace of technological and scientific change and could be seen in such multipurpose federal agencies as the National Forest Service and the National Reclamation Service. As Samuel Hays shows, conservation helped to transform a decentralized, wasteful, and inefficient society into a well-organized, technical, and centrally planned social organization. This "gospel of efficiency" reflected long-range planning and purpose.[106]

In Mead's mind, Australia's state settlement programs illustrated Progressivism's value. Deakin's decentralized programs had wasted money and water. By contrast, closer settlement represented a new social and environmental ethic. Mead extended his staunch belief in conservation and government aid to California, where he believed that the government could carry out works for the common good by shifting away from private enterprise.[107]

Mead hoped that his clarion call for government intervention would revive America's core values. At heart, Mead was a social and agrarian reformer, motivated by his hatred of monopoly and his dream of Jeffersonian rural democracy. A mere 310 individuals or corporations owned more than four million acres of California's farmland. More than three-quarters of publicly held lands in the state had fallen into the hands of railroads, corporations, or individual speculators.[108] Approximately two hundred million acres were owned in tracts of more than four thousand acres, 70 percent of it farmed by tenants—whites, including some from other states, and workers from Mexico, Japan, China, and the Philippines. "We have at one end of the social scale," Mead wrote, "a few rich men who as a rule do not live on their estates, and at the other end either a body of shifting farm laborers or a farm tenantry made up largely of the community."[109] This sort of economy bred a society that lacked effective rural institutions, political stability, and true democracy.[110] Mead believed that only the government could alter existing patterns of landownership and promote social mobility by helping individuals own plots of land. Although he realized that tenancy could not be wholly abolished, he argued that it should be recognized "only as a stepping-stone to ownership."[111]

Mead's visionary rhetoric about state settlement, landownership, and the family farm played directly into the public's awareness of America's threatened rural tradition. The relative proportion of Americans who lived on farms started to decline sharply in the 1860s. The Populist movement's national political failures of the 1890s and 1900s pointed to the decreasing strength of small farmers, agricultural producers, and rural businessmen. By 1910, less than 30 percent of the American population lived on farms, and industrial expansion, urbanization, improved transportation and communication systems, and agricultural mechanization contributed to this decline.

Mead believed that homogenous, middle-class farming communities best represented America's republican values. In this aspect, he embraced parts of the Back to the Land movement, which, in all of its permutations, expressed many Americans' renewed faith in rural life. Stemming mainly from urban unemployment, a dislike of industrialized society, and a fear of declining rural profits, the movement brought together a wide set of ideas: rural nostalgia, humanitarianism, scientific farming, efficient production, and economic self-interest.[112]

Mead's interest in reviving the family farm also betrayed his horror at the number of foreign immigrants flooding the state. Like many other irrigation advocates, he believed in Anglo-European superiority. In the 1890s, the Japanese began to migrate to California as field hands. By 1920, Japanese farmers, who had replaced the excluded Chinese, possessed half a million acres of land in the Central Valley and dominated the state's rice and tomato production. The anti-Asian sentiment that arose in the 1870s and persisted through the 1920s suggested that white Californians felt that their way of life was being threatened by the "yellow peril."[113]

Mead's Australian experience pervaded all aspects of his work in California. In 1915, with the aid of University of California colleagues, he convinced the California legislature to establish the State Commission on Colonization and Rural Credits, which was populated by prominent San Francisco businessmen and chaired by Mead.[114]

A year later, Mead directed an extensive survey to evaluate thirty-two private land settlement schemes. With the aid of county agents, agricultural college professors, rural bank officials, university students, and colonists, the group examined settlements at Carmichael, Chowchilla, Fontana, Imperial Valley, Van Nuys, and Yuba City as well as the Little Landers colonies. The surveyors' report painted a bleak picture of rural life. Many of the districts showed "less social progress" than could be found in European countries.[115] Common workers could not afford to purchase land, and private land companies failed to assist those who did.[116] Worse yet, the average time of repayment for loans ranged from three to eight years, with debt averaging four thousand dollars per settler.[117] By contrast, the colonization systems in Australia, New Zealand, South Africa, and Denmark gave settlers between thirty and sixty-five years to repay loans.[118]

Mead's survey justified state intervention on financial grounds. In California, like Australia, most private land developers, including the Chaffeys, experienced major capital deficits. Mead reassured his American constituents that state control in Australia produced more efficient farming operations and greater profits than private enterprise alone could generate. Comparisons of Mildura's early and later days and of closer settlement of the 1860s with that of the 1920s illustrated that government offered many solutions.[119]

Mead and his supporters tinged their modern Jeffersonian appeal with strong ideas about middle-class respectability and a strong racial philosophy. State settlements supported the preservation of the Anglo-Saxon race by giving men the opportunity to own land. Rural society, Mead predicted, "will either be separated into an upper and lower layer composed of different nationalities, or it will become entirely Asiatic." Helping men to own farms, the title of his popular treatise on rural credits in California, would restore the white family farm as the foundation of American society.[120]

In 1916, Mead submitted his proposal, based on Australia's closer settlement policies, to Governor Johnson, a staunch Progressive reformer who favored centralized planning and conservation. Mead recommended that the state government assist qualified persons in acquiring small, improved farms. The state would purchase a ten-thousand-acre tract selected by scientific experts, who would then divide the land into two hundred farms varying in size depending on soil quality and crops. The land would also hold farm laborer plots and sites for schools, public buildings, and a town. Australia's rural credit system provided the logistical model for financing settlers. Colonists had to pay only 5 percent of the purchase price of the land and 40 percent of the cost of the improvements.[121]

In 1917, state senator Arthur H. Breed, a prominent real estate agent from Oakland, authored the Land Settlement Bill. Modeled after Mead's Victorian Act of 1909, the measure authorized funds for two state-run colonization projects and appropriated $260,000 for the Land Settlement Board, to be chaired by Mead. The board's goals meshed the yeoman farmer ideal with modern, bureaucratic methods. It hoped to prevent the exploitation of naive farmers by private land corporations, to colonize California with a select class of settlers, to demonstrate the effectiveness of expertly planned settlements, and to show the importance of long-term loans at low interest rates.[122]

The Land Settlement Board began its first demonstration in scientific colonization in 1918. It purchased approximately 6,239 acres at Durham, just south of Chico in the Sacramento Valley. The board chose Durham over thirty-nine other candidates for its relatively low cost, its access to the Southern Pacific railway, its rich soil, and its proximity to waters from the Sierra Nevada and Coastal Range. No owner had lived on the land for twenty years.[123]

Experts prepared Durham for settlement. Scientists from the University of California examined the soil, agricultural, and irrigation possibilities. The Division of Horticulture advised on orchard cultivation. Frank Adams, head of the Division of Irrigation Investigations, provided guidance on water use. The Division of Entomology consulted on pests, while the Bureau of Public Roads offered advice on transportation. Architects built homes with electricity and running water. Mead appointed George C. Kreutzer, who had directed an Australian closer settlement, as superintendent of the colony and allocated farm blocks and twenty-six laborer allotments.[124]

Selected from thousands of applicants, the settlers, whom Mead called the New Forty-niners, made 5 percent down payments and agreed to repay the rest of the mortgage over the next twenty years.[125] Criteria for selection of colonists included practical knowledge, character, and possession of a

*The members of the Bigcraft family in front of their allotment at Durham, 1920.*
*Courtesy of the Bancroft Library, University of California, Berkeley*

minimum capital of fifteen hundred dollars.[126] Most settlers came from the United States, but South Africans, Canadians, and Italians also applied. Five thousand people sought farm laborers' allotments in 1921 alone.[127] "Back to the land for me," declared one city engineer who settled in Durham. After inspecting the farms, he crawled into a haystack for the night, just as he had done as a boy.[128] His words reflected the general sentiment at Durham, where 120 families, many previously tenant farmers and representing ten different national backgrounds and all major U.S. regions, settled comfortably into modern rural life.[129]

"Cooperation," a journalist for *Collier's* explained, was the "secret" to Durham's idyllic life.[130] Kreutzer provided practical guidance regarding farm management, livestock and equipment, and cooperatives. Within two years, he had helped settlers form a Stock Breeders' Association and create a central milk station, fair and sports grounds, and a twenty-two-acre grove of oak trees.[131] Mildred Adams, who visited Durham in 1922, first scoffed at

*Making concrete pipe for use in settlers' ditches. Courtesy of the Bancroft Library, University of California, Berkeley*

the settlement's adage, "To rescue for human society the native values of rural life." But after spending a day at Durham, she wondered if she could find an eligible farmer to marry.[132]

In 1919, with Durham established, the California legislature appropriated one million dollars to extend the land settlement program, with preference for war veterans.[133] Mead helped to draft the National Settlement Act under Lane.[134] The Land Settlement Board purchased eighty-seven hundred acres of land in Merced County, in the agricultural heart of the San Joaquin Valley, from the Turlock Irrigation District and private owners. The Southern Pacific and Santa Fe railroads crossed the land, and the settlement abutted the main state highway.[135]

Delhi opened its doors to its first colonists in May 1920. The Land Settlement Board gave settlement preference at Delhi to returning soldiers, a way of solving what Mead predicted would be a major crisis caused by thousands of veterans returning to California with no land on which to settle. Delhi thus continued an American tradition of granting parcels of land to veterans, although few soldiers had the initial capital to enter the settlement.[136] Settlers soon occupied 95 farm laborer allotments of two acres each, 65 small poultry farms, and 234 general farms of thirty acres each. Although the major enterprise at Durham had become dairying, the Delhi settlers raised poultry and grew fruit and vegetables. With the aid of the Department of Agriculture, Delhi residents formed a cooperative for the sale of their fruit.[137]

One writer for *Sunset Magazine* compared the easterner in the 1870s who went West, took up a piece of government land, and became a farmer to the office worker of the 1920s, for whom free land had disappeared. The writer argued that Frederick Jackson Turner's ideas remained relevant thirty years after he posited the closing of the American frontier, which still threatened America's core values of individualism, social mobility, and democracy. The reporter saw the country as occupying no better position than had been the case in 1890 and argued that the "safety valve" of free land had stopped functioning altogether.[138] Mead agreed. With the disappearance of free land, he wrote, "this country entered on a new social and economic era. Free land had furnished an open road to economic independence. This road has been closed."[139]

The colonies also drew worldwide attention. Mead became a household name among irrigation farmers everywhere.[140] Officials from more than forty foreign countries visited the two settlements. Colorado, North Carolina, and South Carolina government officials designed but never launched similar programs. Arizona, Nebraska, Maryland, Washington, Minnesota, and South Dakota began state-settlement projects, although none succeeded. Wyoming likewise considered such a program.[141]

Yet both Delhi and Durham experienced the same economic hardship that had plagued the Chaffeys' Mildura colony thirty years earlier. The Land Settlement Board designed Durham during prosperous times, but Delhi began during a recession that bankrupted thousands of farms and experienced farmers. Although Durham fared better than Delhi, both experienced financial turmoil.[142]

The two colonies floundered for other reasons as well. First, despite the renewed lure of rural life, the state experimented with the colonies at a time when cities were still attracting more people. Back to the Land movements appealed more in rhetoric than in reality. As at Mildura, environmental factors also played a role. Delhi's land required a great deal of leveling and the installation of an extensive underground concrete pipe system for irrigation. Such improvements raised the ultimate cost of the land to four hundred dollars per acre, much higher than colonists had expected. Moreover, the soils in both locales were not as productive as planners had believed, and crop diseases produced low yields. The Delhi settlement had been particularly badly chosen. One of the land's previous owners later admitted to Adams, "We always spoke of that tract of land as one over which the jack rabbits carried their lunches when they passed over it."[143]

The issue of financing also continually resurfaced, particularly in Delhi, where orchards did not bear fruit for several years. Mead repeatedly extended generous loans to keep "marginal" settlers going, to no avail. The first protest at Delhi occurred in 1923 over an appeal for more financial relief. When Mead refused the setters' request, they attacked a portrait of him hanging in the community hall. By 1925, the State of California had become increasingly hostile to the farm settlements and began to withdraw from the field of colonization.[144]

After five years of bitter litigation, the state reached a settlement with the colonists and turned over the management to them. In 1931, the legislature abolished the State Land Settlement Board. In addition to bringing great hardship on its settlers, the experiment in scientific colonization had cost California $2.5 million.[145]

Mead had hoped to create four or five colonies each year in California. But after Durham and Delhi's failures, he admitted to some of his mistakes. He also blamed the depression, a new governor, and irresponsible colonists. State senator C. H. Deuel, editor of the *Chico Record* and an enthusiastic advocate of state land settlement, expressed the romantic impracticality of the colonies: "Ideal in theory the plan failed in California . . . because it was ideal and not practical. . . . It was conceived in the fat times of agriculture and made no provision for the lean years following the war. It assumed that the State as an impersonal agent could not err in values . . . nor that the 'experts' of the State could prove fallible. It attempted to order the lives of a large group of families under a set of blanket rules. . . . It is better, in fact it was necessary," Deuel ruefully concluded, "that the State and the settlers be separated."[146]

Mead's state experts and rural credit system could not take all the blame, however. The same land settlement program succeeded in Australia because of the strong government support it elicited during times of economic prosperity. Despite their lower levels of enthusiasm for Mead's experiments in "state socialism," California officials gave it a trial run, for better or for worse.[147]

Despite the California colonies' failures, Mead held fast to the ideals that had inspired his Australian and American projects: the reclamation of unproductive lands, the revival of rural life, and the application of Progressive ideals to agriculture. Under the right conditions, he believed, technology, expert planning, and government intervention could make rural life more attractive. But under adverse circumstances, planners abandoned the social ideals associated with a modern rural infrastructure.

Mead exported his state-settlement schemes around the world with varying degrees of success into the late 1920s. He continued the work he had started in Victoria in the 1910s in South Australia and New Zealand. In Hawaii, he used land colonization to resettle members of the vanishing Hawaiian race on their land, to offset the threat of Asian immigrants, and to produce a new American class of yeoman farmers. In Palestine, he worked with Zionist authorities to rebuild a Jewish homeland. And in the United States, he wielded the Bureau of Reclamation as a tool to nationalize land settlement. In all of these efforts, Mead drew attention to the importance of economic and scientific settlement, winning massive government funding for his innovative programs. Although his colonization efforts in California failed, they had much merit. Only a decade later, some New Deal reforms would borrow his concepts of state-aided development in yet another effort to invigorate rural life.

Other California engineers similarly endeavored to revitalize society in other parts of the world. In South Africa, these men worked with British entrepreneurs such as Cecil Rhodes in an attempt to transplant Anglo-Saxon technologies and frontier ideals to another gold-rich colonial place. Although these engineers—John Hays Hammond and William Hammond Hall prominent among them—contributed to South Africa's economic growth, they found after several serious mishaps that the region was a far cry from their own. In the end, they helped engender the vast social and racial inequalities that became part and parcel of the white imperialist struggle.

### CHAPTER 4

# Home Is Not So Very Far Away

CIVILIZING THE SOUTH AFRICAN FRONTIER

In December 1895, Cecil Rhodes planned a raid against South Africa's Transvaal government. The plot featured wealthy British mine owners battling alleged injustices by the Transvaal government, a brush with death sentences issued by London's High Court, and public officials' falls from grace. Rhodes's political bungle, called the Jameson Raid after the rebel commander, Dr. Leander Starr Jameson of the British South Africa Company (BSA), gained worldwide notoriety. It resulted in the arrest of sixty-four so-called reformers, including prominent California mining engineer John Hays Hammond and eleven other Americans.

What were American engineers doing in South Africa in 1895? What prompted them to partake in a foolish and dangerous British endeavor? The first question is easy to answer. California engineers converged in South Africa at the end of the nineteenth century for several reasons. In the 1880s, new technical and scientific knowledge expanded hard rock mining activity around the world.[1] A great gold rush began in 1886 along the high, dusty ridge of the Transvaal Republic, the Witwatersrand (Rand). It coincided with Judge Lorenzo Sawyer's injunction against hydraulic mining in California in 1884, which decreased opportunities for local engineers. Many Californians thus joined exploration companies that sent them abroad, some to South Africa. There, they worked for British businessmen, helping to develop the Rand's deep-level mines, Johannesburg's domestic water supplies, and irrigated agriculture in Rhodesia, Cape Colony, and the Transvaal.

Why Californians conspired with British imperialists is more difficult to answer. Engineers saw many similarities between California and South

Africa. White settlers in both places took control over semiarid but fertile, mineral-rich land. They displaced indigenous people and earlier settlers, believing that the new ways of government and culture were better than those that already existed. These parallels suggested to California's engineers that South Africa offered an opportunity to replicate the frontier process that had irrevocably altered California. After establishing ties with British imperialists through the Exploration Company, engineers transported their mining and irrigation expertise to South Africa. They also attempted to transfer the political democracy, social mobility, and free markets that had accompanied the growth of these industries in California. The disastrous results showed the difficulty of transporting American ideals to a colonial region.

The circulation of California's engineers, tools, and ideas helped to build parts of South Africa's economy. Yet these tools came together in ways that belied their original intent and context in California. South Africa's labor practices, racial tensions, weak markets, and poor transportation systems as well as the tenuous British foothold in the Transvaal further impeded Californians' transfers and the region's economic growth. California's engineers found that in the end, their frontier experience differed greatly from the white imperialist struggle in South Africa, where, William Hammond Hall mistakenly observed, "home [seemed] not so very far away."[2]

### The Exploration Company

In 1886, two California mining engineers, Hamilton Smith and Edmund De Crano, solicited the financial backing of London's Rothschild family and founded the Exploration Company. The company grew into a highly successful enterprise that gathered California's engineers into a centralized troupe and sent them to all corners of the earth. The Exploration Company was almost singularly responsible for involving Californians in British ventures in South Africa. This process illustrates the tightly knit web of capital and expertise that linked British entrepreneurs and California engineers in the late nineteenth century.

Smith, founder of the Exploration Company, was born in 1840 on the Kentucky frontier. After making his way to the Pacific Coast when he was almost twenty to work in the mining industry, he rose to become superintendent and chief engineer of the North Bloomfield Gravel Mining Company, Northern California's largest debris-producing mine. Smith soon became one of the state's foremost authorities on hydraulic mining, serving as president of the Miners' Association. Solidly defending mining interests, he played a prominent role in the legal fight between the Sacramento Valley's miners and farmers.[3]

The connection between British capitalists and California engineers formed in the 1870s, when British entrepreneurs looked at California's gold mines as a ripe investment opportunity.[4] In the late 1870s, Baron Edmond de Rothschild, son of the man who had facilitated Benjamin Disraeli's purchase of the Suez Canal in 1875, visited California and invested in North Bloomfield's hydraulic mines. Rothschild was greatly impressed with Smith and hired him to consult on other American and South American gold-mining interests. Smith's success in these ventures elicited Rothschild's lasting support.

In 1886, upon Rothschild's advice and the backing of the Rothschilds and twenty other prominent London businessmen, including A. Mosenthal and Company, a diamond merchant and finance house, and two directors of the BSA, Smith entered into partnership with De Crano. The two engineers formed the Exploration Company to purchase mines and mining rights and act as an agent in their sales. The London-based firm's £20,000 initial capitalization served as an exploring fund and expanded to £300,000 in 1889 and £1,250,000 in 1896.[5]

The Exploration Company served as an import-export agency for capital and expertise. As manager and consulting engineer of the Exploration Company, Smith introduced American mining securities into the British market and obtained British financing for mines in the United States, Mexico, Venezuela, Alaska, and South Africa. Through the company, the Rothschilds held controlling interests in many of Colorado's, Idaho's, and California's major mines, including 40 percent of the Anaconda copper mine. The Exploration Company also formed partnerships with other companies in South Africa and Australia.[6]

The Exploration Company drew heavily on American engineers. By the 1890s, engineers all over the world deemed technical schools in the United States the best. The heaviest demands for Americans came from Mexico, western Australia, and South Africa, where mining played a significant role in national economies. One American engineer even noted "how small the world appears to us Mining Engineers. We meet each other in all parts of the world and take it as an everyday occurrence."[7]

The Exploration Company's activities in South Africa showed American engineers' international prominence. Mark Twain noted during his travels around the world, "South Africa seems to be the heaven of the American scientific mining engineer. He gets the choicest places, and keeps them."[8] Smith, for example, visited South Africa in 1892 and 1895 to examine the Transvaal's deep-level gold mines. His reports on the Rand's promising wealth led Americans to invest in South Africa's mines.[9]

The California engineers who filtered through the Exploration Company to South Africa's gold fields became the "brightest ornaments of American mining and metallurgy" on the Rand.[10] One editor explained the penchant for hiring California engineers in managerial posts as a result of their ability to apply "in one place ideas gathered from another place."[11] Another reporter observed that all of the mining machinery, including hoists and engines, as well as cement, came from America, the preferred market for such equipment and materials.[12] San Francisco's Portland Cement Company, Vulcan Iron Works, California Powder Works, and Union Iron Works supplied the bulk of the materials.[13]

Mining, South Africa's first large industry, owed its growth to imperial expansion. The British Empire assumed control of Cape Colony in the early 1800s, when the port began to serve as a supply station for India-bound ships. Merchants traded wine, wheat, cattle, and sheep. The Cape's strategic importance diminished when the Suez Canal opened in 1869, but the British expressed greater interest in the region as Germany, Portugal, France, and Belgium began to scramble for Africa's riches. Boers (Dutch farmers) unearthed diamonds along the Vaal River north of Cape Colony in the 1870s; a decade later, farmers discovered gold in the Rand, the northern part of the Transvaal.[14]

Although technically a Boer republic governed by Paul Kruger, the Transvaal, along with the other republics north of Cape Colony, fell within what Britain considered its sphere of influence. Cecil Rhodes, prime minister of Cape Colony, director of the BSA, chair of the De Beers diamond cartel, and joint managing director of Consolidated Gold Fields, led the BSA into Boer territory. His desire to create a "foundation of so great a power as hereafter to render wars impossible and promote the best interests of humanity" rested on black labor and the subjugation of Boers.[15]

Californians joined Rhodes's imperial mining projects through association with the Exploration Company. Gardner F. Williams, one of the first California engineers to filter through the company to South Africa, forged the initial ties between Rhodes and his American employees. In 1884, Smith and De Crano hired Williams to take charge of a mining property in the Transvaal. In 1887, after the Transvaal adventure failed, Williams boarded a steamer bound for London and met the thirty-four-year-old Rhodes onboard. During the journey, the two men discussed Rhodes's primary interests in South Africa: the consolidation of his diamond claims in Kimberley, and British expansion to the northern Boer republics of the Orange Free Republic and the Transvaal as well as to Matabeleland, Mashonaland, and Bechuanaland.

Rhodes had gone to South Africa in 1871 to work on his brother's cotton farm but soon acquired a claim in the Kimberley diamond mine and formed a partnership with adjoining claim owner Charles Rudd. In 1880, Rhodes formed the De Beers Mining Company. More than two thousand individual owners of different nationalities shared more than thirty-six hundred separate claims. The cost of mining grew, while the unregulated supply of diamonds fluctuated. Rhodes believed that the only solution to this mismanagement of the mines was to combine all of Kimberley's mines under one company.

In May 1887, Rhodes appointed Williams manager of De Beers, a post he held until the day he returned to San Francisco in 1905. When Williams joined the company, Rhodes faced the problem of how to buy out his competitors.[16] Williams's association with the Exploration Company proved invaluable. The Exploration Company enthusiastically received Rhodes's amalgamation idea and furnished him with £750,000. In turn, Rhodes issued De Beers's shares and paid the Rothschilds a generous commission. Within two days, the Exploration Company had agreed to purchase Rhodes's two largest competitors. This deal linked the Exploration Company and its California engineers to Rhodes's many imperial ventures.[17]

The alliance between Rhodes and the Exploration Company helps to explain Californians' strong presence in South Africa. California engineers went through Smith and De Crano's able hands to Rhodes's gold fields and the Cape Colony government. Once there, they brought some of the battles that had just been fought between the American North and South to South Africa: industrial development, rule of law, free markets, political equality, and civic independence.

### The Great City of Mines: Importing the Tools of Progress

California's engineers came to South Africa with distinct ideas about frontier societies. These engineers perceived themselves as having rapidly transformed California from an unruly place of Indians, Mexicans, and miners into a place of law and order. Hammond, one of the key American mining engineers on the Rand, believed that this frontier process could be replicated in South Africa, which looked "strikingly like the western plains of America."[18] Like California, the Transvaal's mining industry drew together a remarkable assembly of outside capital, labor, and technology. Yet Californians soon found that the mining economy operated under unfamiliar conditions and engendered different social relations than was the case in California.

The gold reef city of Johannesburg and its environs, standing at six thousand feet on top of the Witwatersrand (White Water's Ridge), exhibited vast industrial development. Hall, who consulted for British mining companies in the Transvaal and the Cape Colony government in 1897 and 1899, noted that gold had transformed the barbarous wilderness into civilization. The hoisting works, crushing mills, cyanide and chlorination plants, and machine shops that stretched for miles around Johannesburg reflected the march of progress.[19]

Johannesburg followed San Francisco's path of development. Hordes of gold seekers poured into the South African town almost overnight—a mix of Cornish and Australians, English and Boers, Americans and Germans, Jews and Frenchmen, Italians and Greeks. By 1895, a "great city of mines" had grown up around Market Square, where only twenty years earlier Boer farmers had tended cattle and sheep. The Stock Exchange, high-priced retail shops, banks, mining company offices, gambling, horse racing, and the Rand Club reminded Hall of San Francisco's gold rush days.[20]

Like San Francisco, Johannesburg initially had no infrastructure—only polluted water supplies and dusty, unpaved roads. The town also developed with little adherence to class structure. Although different social and economic classes existed—the English, for example, viewed themselves as superior to the Cornish—Johannesburg was not a rigid class society except with regard to black and white. "No thoughtful man," Hall noted, "could stay long enough in Johannesburg . . . without realizing that he had come upon one of the world's curiosities of social development." Yet like San Francisco, Johannesburg leveled social distinctions only to build them right back up.[21]

With their Western cultural imports in tow, Europeans and Americans arrived as a consequence of the discovery of gold. The Rand contained the world's largest gold field, 170 miles by 100 miles. The Main Reef alone, which protruded as an outcrop, extended for fifteen miles. The gold reefs (known as ledges in America) sloped deeply below the surface in complex formations. Possessing the Rand's riches meant digging deep into the earth, mining on a colossal scale, and marshaling immense quantities of capital and labor.[22]

The Rand presented unique mining problems in the 1880s, but an extraordinary infusion of capital from British, European, and American companies allowed the region to produce enough gold to confirm that metal rather than silver as the world's currency standard. The mobilization of labor also strengthened the mining industry. The Rand had unusual labor practices, although precedents existed elsewhere. Mine owners recruited highly paid, skilled white miners from overseas and quickly promoted them to supervisors over low paid, unskilled African workers. Americans fulfilled most "first rank" posts—mine managers, superintendents, and engineers—and by the 1890s occupied almost 90 percent of the industry's technical managerial positions.[23]

This form of job fragmentation occurred not through the use of sophisticated machinery but rather through the use of large numbers of African workers. In California, Chinese miners occupied the lowest rung of the labor pool; in South Africa, Africans held that status. The opening up of the African interior in the mid-1880s introduced plentiful black labor from as far away as Basutoland, Zululand, Mozambique, and other territories in the north that the BSA had seized. Townships for the black miners sprawled beyond Johannesburg.[24] In 1896, the *Mining and Scientific Press* estimated that forty-five thousand unskilled black natives worked underground in the Rand's gold mines, five thousand Africans were employed in the coal mines at Boksburg, and another fifteen thousand blacks worked in Johannesburg and along the reef.[25] The number of unskilled black workers, who earned approximately fifteen pounds a month (often in alcohol), reached one hundred thousand by the end of the decade. This population declined sharply after the Anglo-Boer War, forcing mine owners to import Chinese labor.[26]

Finally, new technical developments built the industry's infrastructure. Natal provided cheap coal through the 1880s; a decade later, the British developed the East Rand coal field and ran rail lines from the Cape and Delagoa Bay on the east coast. The Macarthur-Forrest cyanide process enabled engineers to process low-grade pyritic ore cheaply. Despite high tariffs, British mine owners imported much of their merchandise and machinery from America, including pumps, cables, and big stamp mills. By 1894, more than two thousand stamp mills crushed 2.75 million tons of ore annually and produced gold worth £7 million, a profit that increased to £82.5 million at the height of the late 1890s boom.[27]

In 1893, Hammond went to London to meet with Barney Barnato, a powerful Randlord, one of the mining entrepreneurs who controlled South Africa's diamond and gold mining industry. After failing to oust Rhodes from Kimberley's diamond fields a decade earlier, Barnato had joined Rhodes and Alfred Beit in the great De Beers Consolidated Company and speculated in gold mines on the Rand. After haggling over Hammond's

*Black and white miners. Courtesy of the Bancroft Library, University of California, Berkeley*

*Tailings wheel. Courtesy of the Bancroft Library, University of California, Berkeley*

*Headgear, Witwatersrand. Courtesy of the Bancroft Library, University of California, Berkeley*

fifty-thousand-dollar consulting fee, the largest of any mining engineer in South Africa, Hammond and his family left San Francisco and settled comfortably into Johannesburg society.[28]

Hammond urged Barnato to consider the possibilities of deep-level mining—the working of ore bodies several thousand feet lower than those commonly worked in South Africa at that time, a tried-and-true method in California's quartz mines. Yet Barnato insisted on using what Hammond considered outdated mining methods, and he resigned after just six months.[29]

After learning of Hammond's resignation, Rhodes offered the American the opportunity to develop Consolidated Gold Fields' deep-level mines at the exorbitant salary of seventy-five thousand dollars a year, plus a share of all profits. Rhodes had previously negotiated with Smith and De Crano regarding various properties and had incorporated Gold Fields with Rudd and Beit. Rhodes hired Hammond as chief consulting engineer of Gold Fields and the BSA, which controlled the mining rights of Mashonaland and Matabeleland.[30]

*Cyanide warls, Witwatersrand. Courtesy of the Bancroft Library, University of California, Berkeley*

Hammond exaggerated his claim that he pioneered deep-level mining on the Rand, yet he certainly accelerated the development of its vast mines, tripling the average monthly shaft-sinking speed and setting new standards of efficiency.[31] He involved Gold Fields in its most ambitious projects, enlisting Gold Fields, together with Wernher, Beit, and Company, a powerful mining house, in Gold Fields Deep. This company eventually controlled all the mining areas of the Central Rand, including the key gold producer in south Johannesburg, Robinson Deep. In 1894, Hammond also staked deep-level claims east of Johannesburg. He developed the Simmer and Jack mine and the Sub Nigel mines, which, along with Robinson Deep, became the world's largest gold mines.

Unlike many cities, Johannesburg lacked a plentiful water supply, a situation that became one of the biggest obstacles to developing the mines. Pumping stations and machinery required water as a power and cooling source, and workers needed water for drinking and bathing. The gold fields stretched more than thirty miles from the Vaal River, the region's only significant waterway. In mid-1888, the Johannesburg Waterworks, Estates, and Exploration Company laid lines from a reservoir in Doornfontein. Other mining companies built small reservoirs, but the mines nevertheless closed down periodically for want of water. Many reservoir embankments burst when the rains finally came.

Beginning in 1895, the area experienced drought. The local waterworks company, a monopolistic enterprise run by Kruger's government, had not properly conserved water from the scarce, seasonal streams. The mines felt the strain, as did Johannesburg's population, which increased by an average of 250 new settlers per day.[32]

Rhodes, who protested that Kruger's officials did not appear to be making any provisions for water, felt that the mines would have to supply their own water. That year, Gold Fields and Werner, Beit, and Company, which together controlled about two-thirds of the producing mines and seven-eighths of the deep-level mining properties on the Central Rand, vowed jointly to undertake a massive water supply plan. Hammond, Beit, Henry C. Perkins (Gold Fields' secretary and Smith's Exploration Company partner and colleague from California's North Bloomfield mines), and Sir Lionel Phillips (president of the Transvaal Chamber of Mines) began to discuss constructing a water supply for the deep-level mines and surrounding residential populations. The men realized that they could never decrease the cost of living, mining, and agriculture until they procured a cheap and abundant water supply.[33]

The partners first proposed a large dam at Vierfontein, directly south of Johannesburg, which would provide water for Robinson Deep. Hammond predicted that an earthwork dam at Vierfontein would quell the flood-waters of the Klip River and store about 1.5 million gallons per day for the mines and Johannesburg's domestic needs. Hammond and Beit submitted the Vierfontein plan to Smith, who was working in America and unavailable to return to the Transvaal. Hammond then invited his cousin, Hall, to undertake the project.[34]

Hall's appointment caused controversy from the start. He first squabbled over his consulting fee.[35] Then, because he had been hired by the Exploration Company with Hammond's influence, he also excited the wrath of its directors. A decade earlier, when Hall served as California's state engineer, Smith and Perkins had viciously fought Hall's debris dams, the stopgap measure to combat the damage caused by hydraulic mining.[36] Blood ties overcame past enmity, however, and Hall agreed to act as consulting hydraulic engineer on all of the water supply projects and waterworks controlled by Gold Fields and Werner, Beit, and Company.[37] Rhodes also hired Hall to consult for the commissioner of public works of Cape Colony, Sir James Sivewright, on irrigation projects.[38]

Hall arrived in Johannesburg in February 1897. By October, he had completed part of the waterworks for the mining houses.[39] He acquired farms owned by Boer farmers, designed pumps for Vierfontein, and bought water rights along the Vaal and Klip Rivers.[40] Hall, however, was an obstinate consultant and involved too many competing interests in his projects. His contentious personality, Kruger's thriving cement monopoly, and British mine owners' competing interests hindered the engineer's work.[41] Hall left the Rand in 1899 under the impression that he had resigned. Wernher, Beit, and Company, by contrast, held fast to the idea that he had been fired.[42] Shortly thereafter, Hall went to Cape Colony to consult for the government there. Not until a few years later did a water board representing all of the Rand municipalities and the mining companies supply water to the town and mines.

California engineers helped to develop the world's largest and most profitable gold mines from what they viewed as a frontier wasteland. Their overall success in transferring their expertise and technology led them to believe that the context in which their industries developed at home would follow: free markets, civic independence, and political liberty. However, the restrictions imposed by Kruger's government on the European-owned mining companies proved otherwise.

*Vierfontein Dam, Wit-*
*watersrand. Courtesy of the*
*Bancroft Library, University*
*of California, Berkeley*

## Our Real Ideal Is Civic Independence:
## Importing the Ideas of Progress

California engineers helped build South Africa's mining economy but had less success in transferring their American ideals to the Transvaal, where Kruger's restraints limited the mining industry's efficiency, curtailed general economic growth, and, according to Hammond and his employers, unjustly curbed Americans' political rights. Hammond thus took action to defend his beliefs in civic independence, laissez-faire, and political equality. Yet his involvement in the Jameson Raid revealed the larger imperative possessed by European and American settlers on the Rand: industrial progress. The raid also hurt Hammond's career, impaired the mining industry, and aggravated already strained relations between England and the Transvaal.

South Africa's colonial environment varied greatly from political conditions in California. One of the major differences related to the existence and strength of previous settlers. In California, white settlers had rapidly decimated the native Indian population, extinguished Mexican land claims, and established statehood. The British thrust in South Africa was much weaker, as the persistence of native black tribes and Boers impeded British settlement. South Africa's black population was much more numerous than the U.S. Native American population and was less easily removed. The Boers, southern Africa's first permanent European settlers, also resisted the British. They had encountered native blacks in the seventeenth century, pushed them northward, and settled into pastoral life, grazing sheep, engaging in small agriculture, and conducting trade. By the nineteenth century, the Boers, who had been living in the Transvaal for two generations under Kruger's oligarchy, resented British rule and the new system it represented.

The British and other Uitlanders (outsiders, or foreigners such as Rhodes and Hammond) encountered many obstacles in the Transvaal. That the Uitlanders were not a unified group complicated matters. The great question of the day in the Transvaal was whether to support Kruger and his republic. Some foreigners, including Rhodes's competitors, hoped to win Kruger's support for their mining endeavors. The largest mining houses, especially Corner House and Gold Fields, increasingly resisted the constraints imposed by Kruger's government; Gold Fields managers in particular wished to see Kruger fall, perceiving him as preying on their gold profits. European-owned gold mines created more than 90 percent of the Transvaal's wealth. But foreigners began to outnumber the Boers in the 1890s, threatening the republic's power from within. Kruger thus took drastic measures to curtail Uitlanders' influence. He placed heavy taxes on the mines' profits and charged high customs duties on imports of mining machinery and manufactured goods from America and England. He also introduced monopoly concessions for the manufacture of dynamite, cement, and alcohol. A Netherlands-owned company also controlled the railways.

The community of Uitlanders had little recourse. Phillips, chair of the Witwatersrand Chamber of Mines, regularly publicized their grievances, to no avail. Since most of the mine owners worked for London-based companies, Kruger considered them "wanderers," not "settlers." While he deprived them of citizenship rights, he nonetheless taxed them exorbitantly. These perceived injustices set Uitlanders on a collision course with Kruger's government.[43]

Many Uitlanders viewed Kruger's restraints as a suffocating check on progress. Hammond condemned the Boer government's penchant for stagnation by describing the Transvaal as "a small unenlightened retrogressive community."[44] The Randlords, the country's most economically powerful men, believed that they lay in the political "clutch, and at the mercy" of "uneducated burghers, scattered about on isolated farms."[45] George Farrar, part of a family firm of mining engineers, viewed the conflict between Boer and Uitlander as a "clash between a pastoral proud and independent race, against a pulsing progressive industrial new population."[46] Above all, Uitlanders denounced the "general inability of the Boers to understand capitalism, industrialization, and progress." Kruger, as founder of the extreme fundamentalist sect Dopper Kerk, was no exception.[47]

Hall also viewed the conflict as pitting stagnation against progress. The Californian met with Kruger to discuss the Transvaal's irrigation possibilities in 1897. Although impressed by the Boer leaders' willingness to build small irrigation projects for his roving shepherds, Hall criticized Kruger's corrupt and stagnant government. Kruger, Hall noted in his unpublished autobiography, "could give our Tammany friends points in their own game." Moreover, Kruger had "gone steadily forward with his programme of retrogression in the face of every . . . pledge to the contrary."[48]

To Hall, the Transvaal's labor system illustrated this backwardness. Along with other Progressive reformers, Hall viewed himself as occupying a middle group between capital and labor and sought to avoid class warfare.[49] Not surprisingly, he applied his beliefs to the unjust labor system he saw unfolding in the Transvaal. In a statement read before the Industrial Commission of Inquiry in Johannesburg in 1897, Hall argued that the problems between the Uitlanders and Kruger's government seemed to involve "national economics," not merely class grievances. By necessity, the Transvaal would have to remodel its political system to satisfy its dominant industries. In Hall's view, the labor question—black or white, skilled or unskilled—reflected Kruger's failure to adjust to the Transvaal's mining industry and posed the most formidable political question of the day. He advocated a new labor system based on capitalist principles—that is, letting the best man win via a bonus system rather than assigning wages simply by trade. Hall's system would apply to all labor groups but improve the blacks' condition in particular. "It would not take [blacks] long to become altogether 'human' with a new wage system," Hall concluded. Yet the Transvaal could solve the labor problem only if the Uitlanders, who controlled the mines but had little say in their cost, became bona fide residents.[50]

The Randlords first plied Kruger with requests for equal political and economic rights. In the fall of 1895, several of the Rand's mining magnates formed a Reform Committee. Led by Colonel Frank Rhodes, Hammond, Phillips, and Farrar, the committee championed what it perceived to be Uitlanders' natural rights. In December, a delegation from the Reform Committee approached Kruger with their demands: full representation in the Transvaal government, proper control of public monies, "true responsibility" to the people, the abolition of railway and public works monopolies, free trade, "pure administration," and equal rights for the English and Dutch languages. According to Hammond, Kruger's only response was, "Their rights. Yes, they'll get them—over my dead body."[51]

The Reform Committee felt justified in taking the drastic step of plotting against Kruger's government.[52] Frank and Cecil Rhodes first gained support from many of the major mining houses, including Gold Fields and Corner House. Cecil Rhodes and Hammond then traveled north to plot their course of action. They met with Jameson, a friend of Rhodes's and administrator of the BSA in Matabeleland, in Rhodesia. The men concluded that only an armed revolt organized in Johannesburg could unseat Kruger and realize Rhodes's vision of a unified South Africa. Hammond returned to the Transvaal, where the four reform leaders formulated a plan to capture Johannesburg and hold the city's women and children captive until Kruger met their demands. They planned to support this uprising with an invasion that Jameson and BSA forces would lead from across the border in Bechuanaland.[53]

Rhodes summoned BSA police from Bechuanaland and Matabeleland, while Jameson stationed his force of six hundred at Pitsani, north of the Transvaal border, to be ready to support the Uitlanders when they rose. Reform leaders had already smuggled English guns into Johannesburg and its environs.[54] Yet Jameson, frustrated by delays on Rhodes's end, decided to move without waiting for the proper signal. On December 29, his men crossed into the Transvaal and headed for Johannesburg. On January 2, as they approached the town, Boer commandos surrounded the invaders and forced them to surrender. Queen Victoria and officials in London, including Joseph Chamberlain, the British colonial secretary, demanded that the reformers surrender their arms. Kruger arrested sixty-four men, including Farrar, Williams, Hammond, Jameson, and eleven other Americans, on charges of sedition and high treason. They were tried in London's High Court. The British government forced Rhodes to resign as prime minister of Cape Colony. Accounts of the embarrassments suffered by the reformers appeared in newspapers around the world and brought Anglo-Boer relations to an all-time low.[55]

The failed raid only increased tensions between Kruger and the British. Rudd, who had not supported the raid, believed that it "could only have been arranged by a madman."[56] Mark Twain, visiting South Africa at the time, also questioned the reformers' motives. "By diligent inquiry in Johannesburg," he wrote, "I found out—apparently—all the details . . . of the quarrel except one—*what they expected to accomplish by an armed uprising.* Nobody seemed to know."[57]

Twain thus struck at the heart of the problem. What had the reformers sought? Had they acted solely to protect their mining interests? Or had they genuinely tried to secure good government for Boer and Uitlander alike? Were not political independence and free markets a part of English settler societies' natural progression? And yet, was not the protection of British and American mining interests also paramount?

The reformers may have possessed different goals, for the men were divided on a number of crucial issues. Cecil Rhodes, for one, never abandoned his plan for the imperial expansion of Britain. The success of the empire in South Africa depended on the continuation of gold mining, which in turn rested on the presence of free markets. Hammond was neither an imperialist nor a staunch British supporter, though he admitted that he and Rhodes were "coconspirators in a political revolution."[58] But Hammond considered the reform movement "in no wise an English movement," particularly since some of the 78 members of the Reform Committee were Americans.[59] In addition, more than 150 Americans had enlisted in the reformers' American George Washington Corps.[60]

Hammond's participation in the raid likely stemmed both from his admiration for Rhodes and from his belief in the ideals behind the raid. Yet he and the other raiders lacked all concern for the details of the exploit. There is no record that Hammond, who naively viewed his role in the raid, considered questions of cost, supplies, and other practicalities.

Hammond's formative years on America's western frontier influenced his perception of how settler societies should develop. Born in San Francisco in 1855, Hammond attended school in the West until he entered the Sheffield Scientific School at Yale University; he subsequently received training at the Royal School of Mines at Freiberg, Germany, before applying his expertise to California's and Mexico's gold fields. His father, Richard Hammond, had arrived in California in 1849 as a major in the U.S. Army and soon became one of San Francisco's most influential citizens, serving as speaker of the state House. Richard Hammond's uncle, Colonel Jack Hays, had earned fame as a Texas Ranger in the Mexican War and notoriety as San Francisco's first sheriff. Both men, who fought for what they believed were their rightful claims to California, served as John Hays Hammond's role models, instilling in him a lifelong respect for the frontier spirit. "One was either a Forty-niner," Hammond's autobiography opens, "or one was not. To the former category belonged the elect who had led the way to fame and fortune."[61]

Although Hammond was a generation removed from California's Forty-niners, his upbringing no doubt inspired his interest in the Uitlanders' cause. The Rand's group of pioneers and conflict between progress and stagnation suggested a replay of California's drama.[62] Yet Hammond might have founded his devotion to the reform movement more on the romantic pioneering spirit than on the reality of his perilous situation. He idealistically thought that the raid would help create an independent Transvaal republic free from British rule. He also believed that the raid would restore mining profits to Gold Fields, though he never quite admitted to that motive.

Hammond invoked patriotic rhetoric to back his views. For him, the reform movement raised the same political issues that had plagued America's first English colonists. In December 1895, before the raid took place, Hammond held a secret meeting in Gold Fields' main office that drew more than 500 Americans. Disturbed by Jameson's intent to carry the British flag into Johannesburg, he secured a flag of the Transvaal and then addressed his audience: "Don't you agree with me that we've now reached the same point as the signers of the Declaration of Independence when they announced that 'it was their right and their duty to throw off a despotic government, and to provide new guards for their future security?' . . . You won't find anything in the Declaration of Independence that limits this principle to latitude or longitude." His rhetoric persuaded 150 Americans to join the George Washington Corps pledge their support to the revolutionary cause of political and economic equality for the Uitlanders and justice for the Boers.[63] Simply put, Hammond wished to see "Republicanism proceeding on the lines of government of the people, by the people, for the people."[64]

Hall, who went to South Africa months after the failed raid, viewed it even more idealistically than his cousin. "Whatever may be our love for Republican institutions and dislike for Imperialism," Hall wrote, "we must not forget that our real ideal [is] civic independence—freedom of the individual, with a liberal form of government . . . self-government."[65] Hall believed that the Boers were "Americans in spirit, though not in name or allegiance."[66] Although the Transvaal was technically a republic with democratically elected officials, the general Boer population had little say in Kruger's regime, which operated more as an ironfisted oligarchy than a democracy.

No one's definition of civic independence included native Africans, whose exclusion fully accorded with American and British notions of progress. Blacks supplied the labor for the mines, which in turn helped modernize the Transvaal. The condition of blacks under imperial rule further suggests that Hammond, despite his bombastic rhetoric, had not been truly concerned about social equality. Rather, rhetoric about civic independence and equality pointed to the more important issue at hand: industrial progress.

Many years later, Hammond expressed his belief that if it had not been for Jameson's rash move, "we would have had a successful and bloodless Revolution." In his opinion, the Union of South Africa could have been formed without the Boer War and without the British claiming the Transvaal and Orange Free State.[67] Hammond viewed his participation in the raid as a civilizing gesture, an attempt to bring justice and good government to a frontier land. Yet the failed raid only exacerbated tensions between Kruger's government and the shamefaced reformers. Hammond left South Africa disappointed. He had spent more than five thousand pounds of his own money on the raid.[68] Even more significantly, he was sentenced to death for high treason, but under pressure from his friends high in the American government, Kruger commuted the sentence and fined Hammond $125,000. "Revolutions," Natalie Harris Hammond admonished her husband as they sailed away from South Africa, "are expensive games to play."[69]

Britain's relationship to the Transvaal proceeded less smoothly than did Hammond's return to the United States. The Jameson Raid debacle catapulted South Africa's relations with Britain onto the international stage and further stoked anti-British feeling among the Boers. The raid brought additional support for Kruger, who overwhelmingly won election to another five-year presidential term in 1898. The revolt also strengthened the alliance between the Transvaal and the Orange Free State. The British Empire stepped in where the private doings of Rhodes, the Randlords, and the Uitlanders had failed. By the late 1890s, European rivalry, particularly from Germany, reached its climax. A solid Transvaal/German alliance, previously supported by many of the American Randlords but feared by the British, threatened the security of Britain's position in southern Africa. The Transvaal constituted the world's largest single source of gold, producing a quarter of the world's annual output. Britain needed control of that gold to retain its position as the world's leading financier.

In 1897, Chamberlain appointed a staunch imperialist, Lord Alfred Milner, to replace Rhodes as Cape governor and high commissioner for South Africa. Milner immediately pressured the Transvaal government to provide the vote for all Uitlanders who had resided there for five years. He brought thousands of British troops to the Cape and Natal to back his demands. Kruger, who realized that the vote would mean transferring political control of his country to the British, refused. Allied with the Orange Free State, Kruger declared war on the British in October 1899.[70]

## Making a Civilised Country: Agriculture and the California Model

As war loomed, Californians proceeded with their engineering projects. Hall, like his younger cousin, applied his perception of frontier development to South Africa. In his unpublished autobiography, he criticized "the richness of South Africa being skimmed from the mining pan," with no part of it going toward developing the country to "support a population after the mining cream had been taken off."[71] Hall viewed irrigated agriculture, which had developed in conjunction with mining in California, as the next stage of South Africa's development. Commercial agriculture, which had yet to be developed on a large scale in South Africa, would carve "a civilised country out of a most savage and barbarous one within the span of one generation's life."[72]

Hall's experience as California's state engineer inspired his plans for Cape Colony, Rhodesia, and the Transvaal, but his vision of fair irrigation laws and systems succeeded in neither California nor South Africa. A great gap existed between Hall's understanding of frontier societies and the reality of their development. Hall's recommendations to the colonial government,

designed to produce an egalitarian and productive society, ultimately supported the imperialist vision of a class- and race-based South Africa.

From the American and European perspective, South Africa was a true
frontier in the 1890s. When one "begins to speculate on the future," wrote
British statesman James Bryce, "his first question is, 'Will these wildernesses
ever become peopled[?] . . . Will South Africa become one of the great producing or manufacturing countries of the world? Will it furnish a great market for European goods?'" Bryce did not ask these questions in vain. As the
new century approached, both Boers and Uitlanders realized that mining
could not sustain South Africa forever. The region had three natural sources
of wealth: agricultural land, pastureland, and minerals. Of the three, agriculture was the least developed.[73]

By the 1850s, Boer and English farms dotted southern Africa. The majority of people in the kingdoms, colonies, and republics lived on a mix-
ture of self-sufficient pastoralism, hunting, and small-scale crop cultivation.
Wool was South Africa's major agricultural export as early as 1840. Commercial farming had not yet expanded from the white-owned grape and wheat
farms of the southwestern Cape and sugar plantations of Natal. Only Cape
Town, Kimberley, and Johannesburg, southern Africa's major cities, generated large enough markets to stimulate commercial food production.[74]

Agriculture, however, required irrigation. South Africa's rainfall ranges
from five inches per year in the Karoo to fifteen inches in the Breeds
River Valley in Cape Colony. "The entire great African peninsula," Hall pre-
dicted, "must be developed by irrigation or not at all." Although parts of
South Africa were not arid, farmers could grow crops in only one-fifth of
Cape Colony and one-third of the Transvaal without perennial irrigation.
Planned irrigation would help create a capitalist base for agriculture and a
new industrial base for the region.[75]

Agriculture would also diversify the risk of mining. As Bryce predicted
in 1900, the gold reefs would not last. But South Africa possessed some fertile ranching and tillage land, and Bryce foresaw Boer farmers and English
settlers building farms and irrigation works. Like Rhodes, Bryce envisioned
a national railway system that would carry agricultural products to distant
markets. Bryce conjectured that in forty years' time, agriculture would have
replaced mining as South Africa's premier industry.[76]

Yet the undeveloped state of agriculture hindered the mining industry.
"You cannot have the best development of the mining industry here in the
midst of a woefully backward agricultural development," Hall argued. Uitlanders had rebelled in part because mining, which furnished more than 90
percent of the Transvaal's wealth, had developed under unjust conditions.

The mines could not operate efficiently until agriculture provided a solid economic base for the region.[77]

The miner, Hall argued before Johannesburg's Industrial Commission in 1898, used the "most perfect mining machinery and appliances" and "the best trained and experienced experts on earth." By contrast, agriculture was crude, its methods wasteful, and its market prices excessively high.[78] Hall believed that agriculture, if coupled with mining, would accelerate South Africa's entry into the industrialized world. Perennial agriculture would also meet the demand created by the growth of a large mining population and decrease the cost of living. English capital would be only too happy to invest in the fertile regions of Britain's African colonies, which would supply delicate fruits such as grapes and oranges to London markets during the English winter and strengthen Britain's commercial foothold in Africa.[79]

As in India, South Africa's irrigation development was tied to the larger process of political and social control. The BSA's projects in Cape Colony and Rhodesia represented the interests of Britain's ruling class: the ability to control nature and natives. Large-scale irrigation, as the Indian, Californian, and Australian models showed, required investment by the state, scientific management, and cheap labor. Using California as a direct model, South Africa's colonial government experimented with this combination in its attempts to irrigate Cape Colony and Rhodesia. Still, agricultural development proceeded slowly. The region lacked the transportation, communication, and market infrastructure that other places boasted and that had made Southern California the hub of the fruit world by the 1890s.

The BSA, a quasi-governmental arm of the state, conducted its first agricultural experiments in Southern Rhodesia. In 1896, two colonial officials formed a department of agriculture to promote the interests of English farmers and to further agricultural progress.[80] The completion of a railway from Cape Colony through Bechuanaland and the protectorate to Buluwayo, which Rhodes also financed and managed, opened potential markets for Rhodesian produce in 1898. Official scientific research, however, did not begin until after the Boer War.[81]

Rhodes also had private agricultural estates in Rhodesia. In July 1897, Hall, acting on his cousin's recommendation, traveled with Rhodes to his namesake protectorate. Hall examined the irrigation and agricultural possibilities on Rhodes's two estates near Bulawayo, Inyanga and Matapo. These lands consisted of a number of leased farms on which English farmers grew deciduous fruits and raised cattle, sheep, and goats. Rhodes hired Hall to build a one-hundred-foot-high dam and large canal at Bulawayo.[82]

Motivated in large part by his racial beliefs, Rhodes carried out these irrigation works at his own expense. He claimed that he wished to show how scientific agriculture could transform the region and its African population. He hoped that his estates would begin the process of transforming the black natives into small capitalist farmers, providing to the rest of South Africa an example of what could be done with the land and people. Rhodes hired hundreds of native workers to cultivate his land.

Hall and Rhodes shared their ideas on world politics during their stay in Bulawayo. One of their discussions concerned the difficulty of developing and governing new regions. Among the greatest problems in South Africa, Rhodes admitted, was "to teach the native how to work and bring him to realize that he must labour." Rhodes advocated compulsory labor on the grounds that America had obligatory school attendance.[83] His farms would give natives work, "supporting them, civilizing them, setting other capitalists and companies a good example."[84] If they refused to work, Rhodes threatened to burn their huts so that they could not return to his land.[85]

Most of the hundreds of natives Rhodes "hired" to farm his estates worked, and he was satisfied to some degree. Yet they showed no signs of becoming small, landowning farmers. They did not care for money and retained their traditional practices of marrying and discarding wives and letting their cattle fall prey to rinderpest. The higher classes of natives also had a great antipathy to labor, which, according to Hall, they considered "derogatory to their dignity."[86]

Rhodes claimed that most Englishmen neither appreciated nor understood the necessity of tying natives to the land. Hall, who had witnessed the disappearance of the Native Americans in California and the racial conflicts leading up to the Civil War, agreed that cheap labor provided the backbone of a new economy. Nonetheless, he questioned the effects of imperialism, a powerful expression of Anglo-European progress, on the natives.

Hall clearly struggled with his beliefs. He had firmly supported the North during the Civil War on antislavery grounds and had refused to attend West Point to avoid living in the South during the war. Yet like others of Anglo-Saxon heritage, he believed in whites' inherent superiority over blacks. At the same time, he acknowledged that progress could occur in South Africa only if the British had a vast labor force. He thus conceded to Rhodes that given the shortage of skilled and unskilled white labor, natives must work. "There was a world of work to do," Hall noted after returning to California, to make South Africa "fit for habitation of, and yield a support to a White population."[87]

Hall viewed the black population with more sympathy than most visitors. He nonetheless felt justified in its conquest, however cruel. The British, he claimed, "have been as far justified in taking the Black man's land in South Africa as were the Americans in despoiling the Red man of his possessions." Yet everywhere Hall turned, he saw hypocrisy, particularly by Americans: "We pride ourselves on enterprise and progressive tendencies. Do we stop to think that these are but secondary qualities, that at the foundation lies ruthlessness?"[88] But he concluded that the subjugation of natives and formation of new social relations were necessary to progress.

While Hall and Rhodes resided at Matapo farm, the two discussed at length the differences between the American and British governments. "You know," Rhodes commented, "I was trying to think . . . of a good simile for the three parts of the dam you were explaining to me . . . but now I have it—the three parts of your Constitutional Government. The front part of the dam enacts the Law that the water shall not go through, the middle impervious clay part executes that law, and the back earthen part judges as to whether the law has been properly enacted and executed, and if it has not been, the whole fabric washes away."[89] The British colonial government lacked these checks. The Jameson Raid, for example, showed how justice applied only to mining magnates. On a horseback trip through Rhodesia, Rhodes admitted to Hall that South African Confederation under the British flag should be effected according to the principles of the American Constitution. Yet as Rhodes's agricultural experiments had shown, progress was a highly race- and class-based concept. The reality of technical progress, compounded with the necessity of cheap labor, precluded the universal application of social and political equality.[90]

Rhodes's Cape Colony fruit farms, which he developed at the same time as his Rhodesian estates, further illustrated the nature of this progress. In the 1890s, Rhodes sent Alfred Mosely, a De Beers diamond merchant and member of the London Exploration Company's board, to California to examine its agricultural techniques. Visitors such as Australia's Alfred Deakin had found much to criticize. Mosely, however, was utterly astonished by what he witnessed in the way of "forethought, energy, and enterprise that is put into the industry there." Engineers and farmers reduced everything to science and left nothing to chance. Mosely projected that "fruit growing at the Cape might be made a large and highly profitable industry if carried out on the Californian lines."[91]

Rhodes, who resided part time at the Groote Schuur estate in Cape Colony, started to develop fruit farms precisely along these lines. Through Hammond, Rhodes hired a young Englishman, H. E. V. Pickstone, who had extensive fruit-growing experience in Southern California. In 1898, Rhodes instructed Pickstone to purchase farms in the Drakenstein Valley in Cape

*William Hammond Hall (standing, at right) and Cecil Rhodes (seated, at right) on top of Matapo.*
*Courtesy of the Bancroft Library, University of California, Berkeley*

Colony with De Beers money. Pickstone bought several farms and another eleven hundred acres in Wellington, Stellenbosch, and Drakenstein. He then became the Rhodes Fruit Farms Company's first manager.[92]

Rhodes modeled his fruit farms after California's and Australia's irrigation colonies. He first lured California farm managers with "superior" education. They built model garden villages with canneries, smithies, mechanics shops, stores, fruit dryers, tile and brick factories, fruit distributing agencies, and central packing houses. They imported the finest machinery from San Francisco and England and bred American farm animals. A short railway ran straight from the port to Rhodes's estate. The English residents on the farms practiced temperance, as in some of Southern California's irrigation colonies.[93]

Most of Southern California's colonies, including Ontario, relied on a hierarchical labor system, with Chinese and Mexican wage workers living outside the settlements' boundaries. Similarly, Rhodes's fruit farms were exclusively white colonies. Factories run by white (Boer) skilled labor and unskilled black labor supported the "white villages." The farms represented a microcosm of a society that British officials wished to duplicate throughout

South Africa—an Anglo-Saxon society built on the toil of native blacks and subjugation of Boers.[94]

Hall's work for the Cape Colony government further supported this vision. He had previously expressed the need to use Africans as cheap labor and advocated domination over the Boers, despite his stated belief that at heart they treasured American values. During the year that he worked for Gold Fields in Johannesburg, Hall established close ties with Sivewright. In 1898, Sivewright hired Hall to provide information concerning the conservation of water and irrigation of lands in Cape Colony. En route from Cape Town to Johannesburg for his mining engagement, Hall spent a month consulting for the Cape government; he then devoted five months of the following year to developing his recommendations into policy.[95]

Hall worked with Sivewright on various public works projects for the colony and consulted for local irrigation commissions. The Department of Public Works had already begun small irrigation schemes along the Vaal River but lacked comprehensive legislation and guidelines for future development.[96] Hall helped Sivewright design an irrigation bill that dealt with the country's floodwaters and riparian rights. He also recommended that a government commission immediately undertake a hydrographic survey of the region's watersheds and irrigable lands.[97]

Hall believed that South Africa's geographical similarities to California allowed the region to repeat "some of the splendid horticultural successes which have been made in California," an assessment supported by Rhodes's fruit farms. The rich soils and favorable formation of the lands for irrigation particularly impressed Hall. Yet southern Africa's two significant rivers, the Orange and its largest tributary, the Vaal, held water only seasonally. By tapping the flows of these rivers during the short wet season and storing floodwaters in reservoirs, Hall calculated that Cape Colony could emulate Southern California's profitable orange- and grape-growing enterprises.[98]

South Africa resembled California in other respects as well. In Hall's words, the "Californian or travelled American feels that he has found another Southern California," particularly in the wetter parts of Cape Colony, where the landscape featured valleys, young orange groves, and vineyards that "bring to mind the best parts of Santa Barbara, Los Angeles, and San Bernardino counties." Hall also noticed native California plants, including Monterey pines. Cape Colony's rich, deep soil reminded Hall of Riverside and Redlands, and he predicted that a stock of "hospitable gentle people," including white Dutch, English, and Americans, could create a yeoman empire out of scattered Boer and English farms. He concluded that these similarities made home feel "not so very far away."[99]

Hall also acknowledged some of the major differences between California and the Cape. First, as he had already noted, the Cape had abundant and cheap labor. The number of native blacks willing to travel hundreds of miles for work eliminated the sorts of chronic labor shortages that plagued California. Fruit could also be transported much more easily in the Cape, despite America's more advanced rail and water network and California's better home markets. Produce grown in the colony traveled straight to the port at Cape Town and thence to England in fewer than fifteen days. The Cape could ripen and market its fruit to meet demand at the English midwinter. In California, farmers carried fruit some distance to a rail depot for shipment across the continent before it was loaded onboard a steamer to cross the ocean. Only careful packing saved wheat, oranges, and grapes from rotting.

Land in South Africa was also comparatively cheap and readily available. Mosely estimated that land suitable for horticulture could be bought in the Cape for one pound per acre, compared to between one hundred and five hundred pounds per acre in California.[100] The concentration of landownership in California meant that small family farmers often were tenants on large estates, whereas in South Africa, those who wanted it—notably, English settlers—had access to plenty of land through the nineteenth century. Although many of the landed estates were large, particularly along the Cape and in sugar-growing Natal, the owners were not rich. One traveler observed that these large owners "excite no envy by their possession of a profitable monopoly."[101]

More important, Hall believed that wheat and fruit production would quell social unrest and address the "poor white question." Some Boers profited from the diamond and gold mines, but most Boers had become restive. Hall thought that the creation of irrigation colonies and irrigation districts, as had been attempted in Southern California and Victoria, would help settle poor farmers and transform them into American-style capitalists.[102] Framed by proper educational institutions and laws, agricultural labor would fix the roving Boer population and the more "intelligent" English settlers on the soil. Permanent agricultural settlement would also make South Africa easier to govern.[103]

Settling farmers on the soil required developing a comprehensive political and scientific irrigation framework. Hall, like his predecessors George Davidson and Deakin, used a comparative approach for his inquiries into South Africa's future. Hall carefully studied the physical conditions, irrigation customs, and water right laws in the Cape before submitting a draft irrigation bill in 1897. He then reviewed these subjects in relation to the best examples in America and Australia: California, Colorado, Montana, New South Wales, and Victoria. Hall based his final report to the Cape Colony government on his recommendations as California's state engineer.[104]

Hall advocated a mix of local and state and public and private works and laws. He recommended that irrigation be undertaken "in a broad way and with a strong hand."[105] Especially in an arid, unsettled region, the government should initiate construction of irrigation and storage works with the idea of ultimately turning them over to local districts. Bearing in mind California's limitations, Hall believed that the colonial government should subsidize the sale of lands, help to finance irrigation works, reduce taxes on land with works designed to increase land value, transport materials on government railways, provide loans for irrigation trusts and associations, and supply engineering assistance. Victoria had learned from California's failures and expended more than three million pounds on national irrigation works and trusts. If private irrigation enterprise failed in America, it would surely fail in countries lacking strong capitalist bases.[106]

Above all, Cape Colony needed "broad and elastic" laws to address the variety of conditions that would make different forms of irrigation enterprise equally desirable: individual works of landowners and farmers, simple associations or syndicates of farmers and landowners, and even irrigation districts. But only time and experience would reveal the best method for irrigation in South Africa. According to Hall, "Irrigation does not freely develop in any country along any single line of enterprise" and was "not a thing to be rushed into lightly or thoughtlessly by either great landowners, capitalists or Governments."[107]

Before policy makers framed local and national water laws, they needed knowledge of existing water claims, use of streams and rivers, rainfall, run-off, stream flow, groundwater, and opportunities for floodwater storage. Hall suggested that Parliament provide for a hydrographic survey of South Africa, even if it proceeded slowly. Cape Colony also needed an administrative framework for irrigation. Hall believed that in accordance with democratic governance, all waters should belong to the people of the country under the stewardship of the Crown. California's experience showed that private ownership of water often fell prey to disastrous litigation. There could be no more certain way of retarding Cape Colony's agriculture than by starting an irrigation "boom" on a basis that produced expensive litigation. Hall believed that special administrative bodies, such as "water courts" or "juries of irrigation," would avoid problems before they arose.[108]

Hall envisioned that officials would then designate water districts according to natural watersheds, estimate the cost of irrigation works for each district, and publish the results in both English and Dutch, a recommendation that hinted at the potentially fruitful alliance between Boer farmers and the colonial government. "The Engineer or irrigation expert who turns a deaf ear to the experiences and the observations of residents and farmers in a country whose irrigation problems he is called upon to solve," Hall warned, "makes a mistake which unfits him for the duty."[109]

Hall, like Hammond, used the American language of progress—civic independence, local rule, and general material advancement—to encourage the colonial government to undertake large-scale irrigation. Yet Hall's recommendations implicitly entailed transforming small Boer farmers into American-style capitalist farmers. Some Boers had, of course, experimented with irrigation, but here Hall revealed his American bias. His strategy required an enlightened conquest of irrigable lands as well domination over future Boer cultivators. By arguing that irrigation required rational, scientific development, Hall provided further justification for imperial rule.[110]

The Boers, the largest group among the white minority in South Africa, posed one of the biggest perceived obstacles to large-scale irrigation. One Australian settler in the Transvaal compared the "progressive" English homestead, with its orderly patches of vegetables and wheat, with the "semi-progressive" Boer farm. Water flowed everywhere, but with no sign of agriculture save a "few patches mealies in low-lying spots" barely tended by the lazy family.[111]

Hall noted the Transvaal's immense agricultural possibilities but like the Australian visitor believed that religion caused the Boers to be "uncompromising in their opposition to real progress." The Boer was a cultivator in name but not in practice. Like the native black tribes, Boers were "little removed from savagery."[112] Their retrogressive, shifting, and unskilled farming methods had no place in the modern world. "It is interesting though sad," Hall commented, "to find how the battles of progress against prejudice and ignorance have to be fought out in every new country." Give the Boers sound knowledge, he continued, and they would "constitute an excellent class of citizens." Instead, they clung to tradition and religion. Kruger's government opposed all efforts to destroy the locusts that periodically razed crops. And unlike Americans and Europeans, whose idea of progress rested on the exploitation of nature, the Boers believed that irrigation mocked God's larger design.[113]

Hall transferred his engineering expertise and irrigation plans, but the colonial environment made his policies difficult to implement. Despite Rhodes's Cape-to-Cairo dream, southern Africa lacked a national railroad system as well as reliable markets and alliances between potential producers, landowners, and public officials. Hall hoped that agriculture would unify South Africa by remedying the social dislocation wrought by British mine owners. Instead, the great chasm between the Boers and the Rand-lords made conflict inevitable.[114] War broke out between Britain and South Africa in the last months of 1899. One year later, the British took control of Johannesburg. Kruger abdicated his presidency and went into exile in Europe.

Rhodes's replacement, Milner, started a rural development program after the war. He attempted to return the Boers to the land, rebuild their ruined homesteads, and restock their farms. Milner also tried to increase British agricultural settlement to balance mining. He encouraged colonists from England, Canada, and New Zealand to become yeoman farmers and act as links between the government and the returning Dutch rural population. But only twelve thousand people took advantage of his offer of free, irrigated land.[115]

The attempt to build a commercial base for irrigated agriculture in the early 1900s failed. Mining, which had also come to a standstill during the war, again began to dominate the economy. Mine owners imported fifty thousand Chinese laborers to replace the Africans who had scattered during the war. The Randlords sent more gold to London, and the Exploration Company continued to send engineers around the globe. Although the high profile of Americans in South Africa diminished, some remained active in the mining industry. Hammond, for example, counseled Gold Fields for many years.[116]

Overall, California engineers failed to implement their model of frontier development in South Africa. Hammond had carried his mining expertise and the American flag to a region hostile to his ideals. Hall had likewise assumed that his irrigation plans for California would work in a place whose agrarian structure, government, and economy differed greatly from his own. Simply put, the tools and ideas that engineers developed in California and exported to South Africa did not adapt well to the colonial environment. Although Californians helped to create a vibrant mining economy, agriculture remained undeveloped. Both industries, far from producing the equitable and prosperous society that Hall and Hammond imagined, furthered imperial goals. In the end, British and American notions of progress revolved around similar ideas about the superiority of Anglo-Saxon settler societies and industrial development. Whether carried out under republican democracy or enlightened imperialism, progress was a universal technical concept that triumphed over historical and cultural differences.

In other parts of the world, particularly the Hawaiian Islands, the growing sugar industry as well as rhetoric about Anglo-Saxon American settlement and the yeoman ideal played out in ways that supported this idea of progress. San Francisco entrepreneur Claus Spreckels's sugar empire facilitated the islands' entry into a global economy, while Elwood Mead's agricultural settlement work attempted to train native Hawaiians in American ways as sugar economics replaced traditional livelihoods and land. But in building this infrastructure and addressing thorny racial issues, their work produced major social, racial, and economic incongruities.

CHAPTER ◆ 5

# Nothing but Commercial Feudalism

CALIFORNIA'S HAWAIIAN EMPIRE

In 1921, Elwood Mead, at the time teaching at the University of California and chairing the state's Land Settlement Board, received a call from George P. Cooke, secretary of the Hawaiian Homes Commission (HHC). Created by an act of the U.S. Congress, the HHC planned to resettle native Hawaiians on government land. By the early 1900s, it had become clear that remedial measures would have to be taken to resurrect the "vanishing" Hawaiian race. HHC leaders believed that settling Hawaiians in agricultural communities on homelands would create a new yeoman class of farmers that would offset the influx of Asian labor. By conforming to modern ideas about work, labor, and capitalism, this new class would also become more "American." Cooke hoped to draw on Mead's expertise in developing suitable land for Hawaiian agricultural settlement. Rehabilitating Hawaiians, however, was easier said than done.

Like California, Australia, and South Africa, the case of Hawaii shows how technical and capitalist growth guided infrastructural development in the late nineteenth and early twentieth century. Pulling Hawaii into a global economy via American imperialism meant growing sugarcane on an unprecedented scale; developing sophisticated irrigation, transportation, and marketing systems; hiring cheap, foreign labor; and dealing with the native Hawaiian question. Also as in California, Australia, and South Africa, Californians facilitated Hawaii's entry into a global economy, but certain incongruities arose. California businessman Claus Spreckels's sugar empire, for example, inextricably tied California's industrial growth to Hawaii's political and economic future and to the islands' destiny: the creation of a poor, urban class of Hawaiians. The HHC and Mead's attempts to resettle Hawaiians on native land on Molokai only further revealed the contradictory effects of progress. Sugar economics catapulted Hawaii into a world economy, but major social inequities resulted.

## Spreckels's Hawaiian Kingdom

On August 24, 1876, the steamer *City of San Francisco* landed at the island of Oahu in the Hawaiian kingdom. The ship carried the message that President Ulysses S. Grant had signed the Reciprocity Treaty between Hawaii and the United States. It also carried Claus Spreckels, a wealthy San Francisco sugar refiner, who was making his first visit to Hawaii. Over the next decade, Spreckels built a sugar empire that tied California's capital, technology, and expertise to Hawaii's commercial and political future. Allying himself with King Kalakaua, Spreckels acquired land and water rights on Maui, hired engineers from California to develop an immense irrigation system, and brought water to his sugar plantation at Spreckelsville. He revolutionized the sugar industry by modernizing production, employing contract labor, designing a railroad system, and marketing the sugar through William G. Irwin and Company. His steamship company, which dominated trade between Hawaii and California, delivered raw sugar to Spreckels's Western Sugar Refinery in San Francisco, bringing the California-Hawaii connection full circle.[1]

The Hawaiian Islands, located in the Pacific Ocean approximately two thousand miles west of San Francisco, comprise eight principal islands ranging in size from 45 to 4,030 square miles. The land surface, made up almost entirely of lava flows, totals approximately 6,500 square miles, a little less than the size of New Jersey. The upper layers of the older lava provide fertile soils, but one-third of the land consists of fresh lava or cliffs, canyons, or ravines. Sugarcane, a giant-stemmed perennial grass that contains sugary juice, grows well on the alluvial flats and lower slopes of the disintegrated lava flows, with fertile grazing land above this belt. Although most of the islands are subtropical, the climate ranges from tropical where land lies below one thousand feet to almost arctic where mountains rise above ten thousand feet. The annual rainfall also varies widely, from four hundred inches to fewer than twenty. A third of the archipelago is arid for almost the entire year, making irrigation a necessity for large-scale sugarcane cultivation.[2]

American interest in Hawaii began in the 1820s, when American whalers created a commercial outpost and brought the Sandwich Islands—named by Captain James Cook for the English Earl of Sandwich—into their orbit. During that decade, Christian missionaries began to settle the islands. Not wishing to invest in the islands until they had implemented Western ideas about law and property rights, Americans converted the Hawaiian kingdom into a constitutional monarchy in 1840. For the next decade, an American held the position of prime minister, a change Americans believed necessary for Hawaii's economic development.[3]

Until the 1860s, primitive technology and limited capital hindered Hawaiian sugar production. Sugar became king only after the Civil War. The 1876 Reciprocity Treaty attracted direct investment by Californians, who viewed the islands as a commercial and strategic gateway to Asia. Proximity to the Pacific Coast and efficient ocean transport made San Francisco sugar's natural market.[4] Mark Twain, who visited the islands as early as 1866, noted, "It is a matter of the utmost importance to the United States that her trade with [Hawaii] should be carefully fostered and augmented. Because — it pays. There can be no better reason than that."[5] Hugh Craig, president of the San Francisco Chamber of Commerce, reflected this belief when he wrote, "The commercial development of the Islands [rests] largely with the merchants and capitalists of San Francisco."[6]

Spreckels, one of San Francisco's largest capitalists, became the islands' most influential investor. The eldest of six children, he was born in 1823 in Hanover, Germany. He attended the local village school, but family circumstances forced him to begin work as a farmhand at age fifteen. To avoid mandatory army service, he immigrated to America, landing in South Carolina in 1846. He worked as a grocery clerk before starting a grocery business. In 1855, he took his wife and children to New York City. Soon thereafter, convinced of the opportunities in gold rush California, he bought a grocery store in San Francisco, and in 1856 he traveled to California via steamer.[7]

Spreckels found San Francisco an open market, with only one sugar manufacturer, the San Francisco Sugar Refinery. The entrepreneurial Spreckels decided to outcompete the competition. In 1864, after he had experimented with the grocery and brewing businesses, Spreckels incorporated the Bay Sugar Refinery, the first of his several successful sugar ventures that linked Hawaiian sugar production to California refineries and markets. In April 1867, Spreckels expanded his operations by founding the California Sugar Refinery.[8]

The California Sugar Refinery soon surpassed its competitors. After studying the process back in Germany, Spreckels designed a new method of making cube sugar that brought product to market within twenty-four hours after the separating process had been started. This innovation marked Spreckels as the most efficient sugar producer on the West Coast. In 1869, one newspaper reported that by producing 250,000 pounds of sugar every ten hours, Spreckels "had bled his opponents, so freely that he foresaw their early retirement."[9]

In 1887, Spreckels reorganized the California Sugar Refinery as the Western Sugar Company. The California and Hawaiian Sugar Refining Corporation at Crockett, which was owned by thirty-three of the fifty-two plantations producing raw sugar in the Hawaiian Islands, provided Spreckels's major competition. Both depended on continuous shipments of raw sugar from Hawaii.[10]

Sugarcane had long been grown on the Hawaiian Islands. When Captain Cook encountered the islands in 1778, he found natives cultivating it in the valleys and lowlands. The first large-scale attempt at sugar cultivation occurred in 1835 at Koloa, on the island of Kauai, by Ladd and Company, a Honolulu firm. But until the 1860s, crude technology hindered production. Planters built small mills run by oxen and horses and used wood rollers and whaling pots to boil the cane's juice. More than 50 percent of the sugarcane area in the islands depends entirely on the artificial distribution of water, but irrigation technology was still in its infancy. By 1875, only ten thousand acres were under sugar cultivation, mainly on Maui.[11]

The Hawaiian Reciprocity Treaty altered the future of Hawaiian sugar production. Approved by the U.S. Senate in 1875 and signed by King Kalakaua the following year, the treaty allowed duty-free trade of some U.S. and Hawaiian products, including sugar and rice. A later version of the treaty ceded to the United States certain rights to Pearl Harbor. The overthrow of the Hawaiian kingdom and the establishment of the provisional government in 1893, followed by the islands' annexation by the United States in 1898, ensured the continuation of these trade benefits.

The Reciprocity Treaty also changed conditions for Hawaiian planters. It devoted full government support to the fledgling sugar industry and pledged aid to growers who wished to develop water supplies and plant sugarcane on marginal land. It also designated Hawaii as a one-crop economy. Without government support, fewer foreigners, including Spreckels, would have looked to Hawaii as an opportunity for investment.[12] "San Francisco," one businessman wrote, "can justly expect to reap a large proportion of the benefit derived from the development of the resources of Hawaii, while the islands themselves will offer a most tempting field for the investment of idle capital, with absolute security and assured and large returns."[13]

Spreckels had "idle" capital awaiting just an opportunity, even though he initially opposed the Reciprocity Treaty in the belief that lowering or removing tariffs on sugar would hurt America's refining industry and end his profitable business. Spreckels thus acted to protect his interests. He traveled to the Hawaiian Islands and contracted with the Hawaiian planters for more than half of the year's sugar production—about seven thousand tons—before the treaty went into effect. His dealings with Hawaiian planters gave him control over the majority of the current Hawaiian sugar crop but did not guarantee his future profits. Spreckels thus initiated the first link in a chain of operations that made him both the largest producer of sugarcane in Hawaii and the largest refiner on the West Coast.[14]

In 1878, two years after his first visit, Spreckels returned to the islands and started looking for land on which to develop his own sugarcane plantation. "I am a planter as well as a manufacturer," he boasted in a San Francisco journal, "and look on the Hawaiian group as another Louisiana."[15] He first needed to acquire land. According to Hawaii's ancient land laws, land belonged to conquering chiefs and their constituents and was revocable. During the reign of King Kamehameha III (1825–54), ways of distributing and owning land changed as newcomers, mostly white missionaries or American planters, sought land. The great *mahele* (division) of 1848 abolished Hawaii's feudal land system and divided lands among the king, government, and people. By 1850, Crown lands could be leased or privately purchased. Many Hawaiians lost their land, and foreigners such as Spreckels started to buy land outright.[16]

Spreckels found Maui, Hawaii's second-largest island, an ideal region for a plantation. Maui spans 760 square miles (400,000 acres) and consists of two circular volcanic groups of mountains. Sugarcane grows well in the island's volcanic red dirt and can be planted in almost any month of the year, since frost never occurs. Plantations had already developed around the mountain bases. Samuel Alexander and Henry Baldwin, whose East Maui Irrigation Company constructed the island's first major irrigation channel, the Hamakua Ditch, in 1878, founded the first large plantation in East Maui. The ditch, which cost eighty thousand dollars and harnessed five large streams, spanned seventeen miles and cut through heavily forested mountains before it reached the cane fields. In West Maui, James Robinson and Company, Thomas Cummins, J. Fuller, and agent C. Brewer and Company had organized the Wailuku Sugar Company in 1862.[17] Charles Nordhoff, one of the American West's famous boosters, marveled that "everywhere [on Maui] the cane seems to thrive, and undoubtedly it is the one product of the islands which succeeds."[18] Spreckels planned to outdo both Alexander and Baldwin and the Wailuku Sugar Company. But two obstacles stood in his way: he owned neither land nor water rights on Maui.

Alexander, an engineer who had constructed irrigation ditches throughout Hawaii, pointed out to Spreckels that the fertile plains of central Maui, a broad stretch of land located between the Haleakala Mountain and the West Maui mountain, needed only water to become productive. In 1878, Spreckels bought an interest in sixteen thousand acres of the Waikapu Commons and leased twenty-four thousand adjacent acres of Wailuku Commons Crown Lands for one thousand dollars a year.

Alliances with King Kalakaua and with Walter Gibson, a former missionary who had carved out an empire on Lanai, were crucial to Spreckels's acquisition of these Crown lands and his eventual purchase of Princess Ruth's lands. Kalakaua and Gibson, who was elected to the Hawaiian legislature in 1878, instituted a regime of political corruption. Realizing that this duo could aid his economic ambitions, Spreckels quickly befriended them and lent Kalakaua money. In turn, the San Franciscan received political favors that furthered his goal of building the island's largest sugar plantation. Over the next few years, with the aid of Kalakaua and Gibson, Spreckels slowly increased his holdings and controlled forty thousand acres by 1882. In a corrupt deal, the legislature later granted Crown lands to Spreckels in fee, which allowed him to develop his plantation without having to worry about land titles or his lease.[19]

Spreckels also applied to the government for water rights. Whereas a mix of prior appropriation and riparian rights (with no clear legal precedent for either) characterized water law in California, in Hawaii, the kingdom and people owned all water rights. Moreover, under traditional Hawaiian practices, most water rights adhered to the land, although water could be leased or sold separate from the land itself subject to other vested rights. Riparian rights did not exist except for domestic purposes until 1893, when English common law was adopted.[20]

These conditions did not deter Spreckels from fulfilling his quest for water. He approached Kalakaua and asked for the desired water rights. Spreckels's request was initially rejected, but the king, already in debt to the Californian and dissatisfied with the cabinet, dismissed its members and replaced them with politicians more sympathetic to Spreckels's enterprise. Spreckels reciprocated with a gift of ten thousand dollars to the king and a loan of forty thousand dollars to the government. Not surprisingly, the new cabinet granted the desired water privileges to Spreckels for thirty years at five hundred dollars per year.[21]

The king rationalized this "gift" to Spreckels by claiming that the Hawaiian government could not incur the expense of undertaking major water projects, leaving the task to Spreckels and other private capitalists. Kalakaua also argued that Spreckels's irrigation enterprise would serve "the agricultural interests and the general welfare of the Hawaiian Kingdom."[22]

Years later, one visitor to Hawaii noted that "the outstanding thing about Hawaiian irrigation works is that they are all privately financed."[23] Still, Spreckels's water deal gave him a few lasting and not altogether flattering monikers, including the "sugar king of Hawaii," the "power behind the throne," and the "maker and breaker of cabinets."[24]

After he had acquired the necessary land and water rights, Spreckels incorporated his holdings. In 1878, together with Herman Schussler, chief engineer of California's notorious water monopoly, Spring Valley Water Works, and San Francisco capitalists William F. Babcock, president of Spring Valley; Frederick F. Low; and Herman Bendel, Spreckels organized the Hawaiian Commercial Company. Spreckels owned the majority of the stock and served as managing director. In 1880, Spreckels formed a partnership with William G. Irwin, Irwin and Company, that became one of the leading sugar agencies in the kingdom and served as financier, merchant, exporter, importer, and marketer of Hawaiian Commercial Company sugar.[25] Irwin and Company also directly controlled one-third of the islands' sugar crop and purchased the remaining two-thirds from other planters. When asked by a reporter if his company monopolized the islands' sugar production, Irwin bragged, "We control almost the entire crop."[26]

In 1882, the Hawaiian Commercial Company reorganized as the Hawaiian Commercial and Sugar Company, a California corporation managed by Irwin and Company that joined the islands' five other leading producers and marketers of sugar: C. Brewer and Company, Davies and Company, Castle and Cooke, Alexander and Baldwin, and Hackfield and Company. These companies (three of which were American) eventually evolved into the Big Five, which owned or controlled most of Hawaii's land, plantations, water, power, mills, labor, transportation, and refineries as well as banks and local politics.[27]

Spreckels's next step entailed building an extensive irrigation system that would carry water to his Maui land. Spreckels redefined large-scale irrigation in Hawaii, eclipsing Alexander and Baldwin's earlier projects in size and scope. Schussler surveyed the region for Spreckels with the understanding that they could enter government lands to build and maintain a large ditch system. Schussler then employed white and Chinese laborers and began to bring water from the rivers on the wetter side of the island to the drier, western side. By the fall of 1878, more than four hundred men, one hundred oxen, and sixty mules were working on the project, with pipe, lumber, and dynamite from San Francisco. The completed irrigation system boasted sixty-five miles of canals and tunnels and crossed thirty-one large ravines with iron pipes.[28]

Spreckels's major ditch was the Haiku Ditch, constructed by Japanese workers using hand drills and requiring a capital outlay of five hundred thousand dollars, the largest amount spent on any irrigation project in the Hawaiian Islands to date. The ditch spanned thirty miles and delivered fifty million gallons of water daily, irrigating twenty times as much land as had previously been irrigated.[29] In 1880, when William Hammond Hall was experimenting with debris dams along California's rivers, the *San Francisco Commercial Herald* reported that Maui's dry plains had become a "mine of wealth, a bonanza, pouring out its ceaseless floods of treasure." Capital, the reporter continued, "went out into the wilderness, and made the desert blossom into cultivated fields, and smiling gardens and happy homes."[30]

The Haiku Ditch rivaled India's great undertakings and surpassed the size of any ditch that had been constructed in the American West. San Francisco city engineer Michael M. O'Shaughnessy compared Spreckels and Schussler to the Mormon settlers of Utah, who under church leadership had constructed some of the West's most sophisticated irrigation systems. "By no other people," he wrote, "has so much enterprise been displayed, and so many sacrifices made in developing a non-productive country into one of pronounced prosperity."[31]

The Haiku Ditch formed only one component of this new enterprise. Spreckels revolutionized Hawaiian sugar production in every way. His plantation and company town, Spreckelsville, became a successful experiment in agricultural and industrial capitalism. The "largest sugar estate in the world" in 1892, Spreckelsville spread over forty thousand acres, with half of the twenty-five thousand acres of good cane land under cultivation.[32] Sugarcane planting started soon after the completion of the irrigation system. Steam plows, the first used in the Hawaiian Islands, turned the ground. Within eighteen months, the crop matured and could be processed.[33]

In anticipation of high yields, Spreckels employed two men from San Francisco's Risdon Iron Works to design a mill able to process twenty tons of sugar each ten-hour day. The mill used five rollers instead of the customary three to extract juice from the cane. Electric lighting, the first of its kind on Maui, facilitated plantation operations. Within ten years, Spreckels added three more mills. The entire sugar production process was almost completely mechanized. Cane was hauled to the plantation mill, where a machine unloaded sugar from the cars. It then passed through a crusher, which extracted the juice and tested for sugar and juice content. The juice was then treated with milk of lime and heated, put into settling tanks, and reduced to syrup in a filtering process. After being boiled and mixed with molasses, the syrup went through centrifugal machines that left grains of dried sugar in the vacuum pan. The raw sugar was then packed and shipped to Spreckels's Western Sugar Refinery in San Francisco.[34]

Spreckels's innovations in transportation also made Hawaiian Commercial and Sugar and Spreckelsville a model plantation system. Plantation transportation had previously consisted of roads traveled by mules and oxen. To reduce the costs of transporting sugarcane to the mills, foreign laborers constructed a narrow-gauge railroad through Spreckelsville. It had two locomotives, one hauling twenty cars and the other thirty. Each car could transport three thousand pounds of cane from the fields to the mills. By 1881, twenty miles of iron track were completed. The plantation railroad connected with the Kahului Railroad, a corporation whose stock was controlled by Hawaiian Commercial and Sugar. This rail line transported the processed sugar to Maui's major port, Kahului. There, Spreckels owned landings, storehouses, and a general dry goods store. By 1885, Spreckelsville had forty-three miles of railroad, four engines, and 498 cars for hauling cane.[35]

Although Spreckels commanded sugarcane production on the Hawaiian Islands, the refineries and markets lay elsewhere. At the end of 1881, Spreckels and his sons organized the Oceanic Steamship Company. By 1883, the Spreckels line operated two steamships and nine sailing vessels, connecting Hawaiian sugar to its markets in America, Australia, and New Zealand. The 247-ton schooner *Claus Spreckels* could make the trip from San Francisco to Hawaii in nine days. With this rapid service, the Oceanic Steamship Company effectively monopolized Hawaiian sugar traffic throughout the world.[36]

Modern irrigation technology, mechanized production, and efficient rail and ocean transport expanded the Spreckelsville plantation's sugar output. The plantation, generally a large farm that produced one or two crops by unskilled labor with a low standard of living, could not have survived without worker relations predicated on cheap, reliable labor. The large size of the farm isolated workers from other employment possibilities and produced a hierarchy ranging from plantation manager at the top to unskilled laborer at the bottom. "The sugar industry and the plantation system have grown up simultaneously," one Hawaii resident commented.[37]

As in California, labor was scarce in Hawaii. Disease and introduced commodities, including alcohol, destroyed the native economy, which was based on subsistence cultivation of taro, yams, plantains, and bananas as well as some interisland trade. The combination of pathogens and the stress of outside encroachment devastated the native population. Between 1778 and 1876, the Hawaiian population decreased from about three hundred

*Waialua Wiel, Maui, from the rear, showing the plantation railroad and Japanese laborers. Courtesy of the Bancroft Library, University of California, Berkeley*

thousand to fifty thousand.[38] Planters tried to secure steady work from Hawaiians, but the Hawaiian "living in his valley on his kuliana or small holding," Nordhoff observed, "leads an extremely irregular life. [He] sleeps a good deal during the day, and has much idle time on his hands."[39] As Hawaiians' numbers dwindled and those remaining increasingly resented working for whites, planters looked to other sources. Twain noted that, "the Kanaka race is passing away. Cheap labor had to be procured by some means or other."[40]

Because they could not rely on the free market and slavery would never curry public favor, planters sought contract labor. The voluntary contract labor system, the engine of Hawaiian sugar plantations, was seen as benefiting all parties. Although reports described the Hawaiian Islands as "a bonanza for these immigrants," planters, who advanced money for the laborers' passage and contracted them to work for a certain length of time, reaped greater benefits. The system protected the sugar planters against strikes and assured workers of regular employment. Although a few labor revolts occurred on the sugar plantations, the contract labor system mostly sidestepped the late-nineteenth-century battles between industrialists and workers that were being fought elsewhere in the world.[41]

The Royal Hawaiian Agricultural Society, founded in 1850 to represent planters' interests, and the Bureau of Immigration, formed in 1864, initiated the immigration of agricultural labor from China, Korea, Japan, the mainland United States, Puerto Rico, Portugal, Germany, Norway, and Oceania. Between 1852 and the end of the century, foreign immigration steadily increased. First to arrive were the Chinese; when their numbers proved inadequate even after the legislature amended its immigration laws, the government began to encourage Japanese immigration in the 1880s. Portuguese, Germans, Norwegians, and Filipinos followed. After the Reciprocity Treaty, more than fifty-five thousand unskilled immigrant laborers entered Hawaii.[42]

The plantation system could not have succeeded without this cheap labor supply and the division of labor it allowed. In Spreckelsville, workers were divided into five camps, each headed by a foreman.[43] Chinese labored in the fields, while Japanese workers constructed irrigation ditches. Each group had a particular skill, as this division of labor suggests. Twain noted that Chinese labor, for example, relieved white men's hands of the "hardest and most exhausting drudgery—drudgery which neither intelligence nor education are required to fit a man for."[44] The Hawaiians were seen as good teamsters—that is, when they would work—and the Chinese and Japanese as efficient factory and field workers.[45]

Sugar plantations nevertheless experienced chronic labor shortages. Many plantation owners did not reside on the land they owned, breeding feelings of tenancy. "Managers and the better paid employees are comfortably housed," Lorrin A. Thurston told the Social Science Club of Honolulu in 1906, "but the great majority of the laborers are sheltered in 'quarters,' in 'barracks,' . . . so small that there can be little privacy."[46] Rare was the plantation of "neatly whitewashed laborers' houses, surrounded with vegetable gardens."[47] In some cases, planters bribed laborers with houses and small tracts of land, but planters generally believed that they could not afford to raise wages because of competition from other sugar-producing countries, including Cuba and Java, where labor was cheap. Laborers earned an average of about one hundred dollars a year, "just about what it used to cost to board and clothe and doctor a negro," Twain commented.[48]

Squalid working conditions prompted many laborers to leave plantations after their contracts expired and seek better employment in the cities or acquire land of their own, as some Japanese did. Thurston recommended that if planters wished to keep foreign workers, each needed to receive a home, a longer term of contract, and a percentage of profit. "Sugar is still, and will long continue to be, King in Hawaii," he argued, "but it will take breadth of spirit and progressive intelligence in handling the labor question, as well as energy, to sustain it in that position."[49] After visiting the islands, Frederick H. Newell, director of the U.S. Reclamation Service, agreed. The decreasing number of men willing to work in the fields, compounded by the increasing demand for labor, posed "an almost impassable barrier" to the lifeblood of the sugar industry.[50] Hawaii's dependence on a single industry, which, in turn, rested on the existence of a protective tariff and continuous influx of foreign labor, did not bode well for the islands' continued prosperity.

The issue of contract labor raised questions on the national, territorial, and personal levels. "The labor question," Spreckels wrote in 1893, "is the all important one and constitutes my only objection to annexation."[51] Spreckels, who saw contract labor as the only viable system, believed that whites could not and Hawaiians would not work in the cane fields. If planters could not hire cheap labor, the plantations would close. Spreckels and the Big Five felt threatened by official American involvement in Hawaiian affairs.[52] Spreckels thus advocated two unpopular positions—that either the monarchy be restored or California annex Hawaii.[53] Other observers saw contract labor as a personal issue, degrading workers, keeping them from landownership, making them migratory, and ultimately turning many into urban wage workers.

*Workers picking sugarcane.*
*Courtesy of the Bancroft Library,*
*University of California, Berkeley*

At the same time, the presence of contract or migratory workers, who formed the majority of Hawaii's population by the 1880s, produced what Newell called "an artificial condition of social and business life not conducive to settlement by whites."[54] For example, a party of Americans sailed from San Francisco to work shares for Spreckels but did not own the land they worked.[55] But, according to Newell, every white worker "must have a home and ownership of a small tract of land if they are to be a stable and conservative force in government."[56] The contract labor system stymied this ideal of a white agricultural class, intermediate between the corporations owning sugar plantations and the landless, migratory laborers. Newell recommended dividing up the public lands as soon as the leases ran out and distributing small tracts to white citizens.

Newell, like others, saw a bleak future for the small farmer in Hawaii. "Can the pioneer white farmer," he asked, "get a foothold on the islands? Will he bring his family to a community composed largely of Asiatics . . . ? Will he be content to labor in his field when most men of his race now act as overseers or employers and where coolies or peasants are hired for all manual work? . . . It is evident that unless he can . . . maintain the standard of living and the social and civic ideals of the Republic, there is little to be gained by attempts to merely increase population."[57] Every possible effort, Newell concluded, "should be made by public and private interests to put upon the land the best obtainable men, who will live upon small farms, cultivate the soil and become independent, self-supporting citizens."[58] Contradictions marred Newell's thinking. Californians in particular wished to bring Hawaii into their economic orbit, and the large sugar plantations fulfilled this vision. Ironically, however, sugar production depended on the existence of contract labor, which most mainland Americans decried. But white settlement could not proceed in its yeoman ideal form within the existing plantation system.

Newell's dream of a society of small landholders thus did not mesh with Spreckels's Hawaiian sugar empire. By 1885, Spreckels had reached the peak of his power in Hawaii. He was the islands' largest sugar producer and controlled almost the entire island crop. He dominated the trade between the islands and the Pacific Coast and exerted powerful political clout.[59]

Spreckels and his San Francisco partners and engineers had reengineered Maui's social, economic, political, and environmental landscape, neatly tying it to San Francisco's commercial interests. His innovations formed many links in a vast industrial chain: the Spreckelsville plantation, with its irrigation ditches, modern production methods, and contract labor; the sugar agency of Irwin and Company, which began to dominate the marketing of Hawaiian sugar in the 1880s; and the Oceanic Steamship Company, the chief shipper of Hawaiian sugar between the 1880s and 1890s. The last link in this great industrial chain was Spreckels's sugar refineries.

By the late 1870s, Spreckels's Hawaiian plantation assured his refinery a steady stream of raw sugar. In 1881, he relocated his refinery to six square blocks in San Francisco, adjacent to the Southern Pacific Railroad tracks and San Francisco Bay. The new twelve-story facility employed five hundred men, with an output of nearly one hundred thousand tons of sugar annually. It had a main refinery, charcoal house, warehouse and melting room, boiler room, engine room, machine shop, coopers' shop, coal sheds, wharf, and residences for married employees. Throughout the decade, Spreckels patented new methods of making sugar—cube sugar, crushed sugar, powdered sugar, fine crushed sugar, and dry granulated sugar.[60]

When he subsequently reorganized the California Sugar Refinery as the Western Sugar Company, Spreckels focused on modern techniques and scientific management, as he had on his Hawaiian plantation. Raw Hawaiian sugar passed from one of Spreckels's schooners over the San Francisco wharf to the warehouse. After being processed, rail and steam power then transported the refined sugar to major ports around the world.[61]

Unlike the Hawaiian plantation, however, the Western Sugar Refinery relied exclusively on white labor. Since the Exclusion Acts of the 1880s limited Chinese immigration to the United States, Spreckels employed German, Slovenian, Irish, Swedish and Danish workers. "The success of this sugar refinery," one newly minted American, a German native, boasted, "is one of which Californians may well feel proud, and it has been built up without the aid of the cheap muscle of a single Chinaman."[62]

By 1886, with his California refinery thriving, Spreckels's Hawaiian power started to ebb. Rebelling against Spreckels's monopoly, planters formed their own company, the California and Hawaiian Sugar Refinery Corporation, which located its headquarters in San Francisco and its refinery at Crockett. It provided major competition to Spreckels's monopoly and ultimately became one of the world's largest refineries.[63]

Spreckels broke with Kalakaua in 1886 over the negotiation of a two-million-dollar loan. A few years later, management problems and the deterioration of facilities at the Spreckelsville plantation weakened his grasp on Hawaiian sugar. In 1898, two notable events loosened his control on the entire industry. First, Spreckels's earlier fear that Hawaiian annexation would bring about the demise of the contract labor system throughout Hawaii was finally realized. Second, Spreckels lost control of the Hawaiian Commercial and Sugar Company to Alexander and Baldwin. Sugar production expanded under its new owners, but the clout that Spreckels had enjoyed in the Hawaiian Islands for almost a decade ended.[64]

Spreckels eventually abandoned Hawaiian sugar for California-grown beet sugar, but Californians continued to play prominent roles in developing the Hawaiian Islands. In the early 1900s, James D. Schuyler, who had served as California's assistant state engineer under William Hammond Hall, helped a colony of California fruit growers develop their water resources on Oahu. He also consulted for the Honolulu Water Works, reported on the water supply for irrigation on the Honouliuli and Kahuku Ranchos on Oahu, and worked on many other irrigation projects throughout the islands.[65] Joseph B. Lippincott, the U.S. Reclamation Service officer involved in developing a federal reclamation project at Yuma, California, and assistant chief engineer for the Bureau of the Los Angeles Aqueduct, brought water to the lands of the Oahu Sugar Company in the early 1910s.[66] O'Shaughnessy completed the largest aqueduct on Maui, the Koolau Ditch, which discharged water into Hawaiian Commercial and Sugar's earlier ditches, and designed other large-scale irrigation projects throughout the islands.[67]

Starting in the 1870s, sugar economics held a prominent place in Hawaiian-Californian relations and securely tied together their futures. Despite the growth in material production in Hawaii, the sugar industry engendered vast social inequities and raised debate regarding the future of white—and Hawaiian—settlement in the islands.

### Resettling the Hawaiians: The Molokai Experiment

In December 1921, Cooke, a millionaire rancher on Molokai and a former territorial senator, wrote to Mead, "I have been deeply interested in your book, *Helping Men Own Farms*. I firmly believe that the industrial development of Hawaii, can be put on a more permanent basis, and one that will be more American."[68] The division and loss of native Hawaiian lands at the behest of the sugar industry as well as the decline of traditional native livelihoods had created a class of poor, urban Hawaiians. As in California, corporate agriculture sent people to the cities as wage laborers. While Cooke mourned the Hawaiians' lost way of life, he also realized that the new era of Hawaiian annexation entailed creating a society that conformed to modern American values. Settling Hawaiians on government-owned homesteads offered a way simultaneously to Americanize the islands, preserve the Hawaiian race, and idealize an old, subsistence form of Hawaiian economy while implementing commercial farm ventures. As the HHC's experiments on Molokai show, HHC home building, which used race as a motivating factor, proved more powerful in rhetoric than in reality.

At the time of annexation, the Republic of Hawaii had ceded roughly 1.8 million acres of land to the United States. Given the nation's interest in sugar growing, returning Hawaiians to the land was a more popular sentiment than returning land to the Hawaiians.[69] The HHC, in the spirit of other land settlement programs around the world in the 1920s, intended to help native Hawaiians regain possession of Hawaiian kingdom government and Crown lands. "Back to the land" was seen as a panacea for the social disorganization of the Hawaiian people. Once settled on government land, the HHC hoped that Hawaiians would make a better living than as wage workers and tenement dwellers in Honolulu and Hilo.[70]

Like Mead's land settlement plans in California and Australia, the HHC fell short of its goals. The HHC Act of 1920 was not inspired exclusively by humanitarian motives. Rather, it was designed as a measure to assure stability and control over the public lands through clauses that undermined native Hawaiians' power. Fearing the loss of productive public lands leased from the government, large sugar planters inserted a provision into the HHC Act that excluded the most fertile and commercially valuable land from settlement. Most of the two hundred thousand acres set aside for the HHC's first experiments in homesteading was unproductive land on Molokai.[71]

Controversy surrounded the definition of "Hawaiian," further darkening the prospects for Hawaiian rehabilitation. An early bill stated that people with ¹⁄₃₂ Hawaiian blood (that is, with one Hawaiian great-great-great grandparent) could apply for land. But the final act restricted eligibility only to people who were at least half Hawaiian, dramatically decreasing the potential number of applicants. It also confused traditional Hawaiian understandings of rank and genealogy. But the HHC Act also stated that 30 percent of the annual receipts from the leasing of sugar land would be loaned to Hawaiian natives for homes, livestock, and improvements on leased lands. Yet the final version of the act also decreased the duration of the leases from 999 to 99 years. Moreover, rehabilitating approximately forty thousand full- and half-blooded Hawaiians could not be accomplished with the HHC's allotted land and funds. What was conceived as a Hawaiian rehabilitation scheme became little more than a way to prevent the breakup of large sugar estates as well as a weak attempt to transform Hawaiians into American-style farmers without giving them property.[72]

The case for Hawaiian rehabilitation was made as early as 1888, when the

Reverend S. E. Bishop read a paper to the Honolulu Social Science Association, "Why Are the Hawaiians Dying Out?" The reason was not, Bishop explained, a case of Darwin's survival of the fittest. Rather, the Hawaiians' decline resulted from what whites had brought to Hawaii, including alcohol and diseases such as measles, smallpox, and leprosy. "The Hawaiian Race is one that is well worth saving," the reverend believed. They "are a noble race of men . . . manly, courageous, enterprising, cordial, generous, unselfish." He concluded that they could be saved by returning them to the soil.[73]

By the early 1900s, conditions had worsened for native Hawaiians. Sugar interests altered traditional landownership and land tenure patterns, threatening Hawaiians' subsistence taro-growing economy. In 1909, whites controlled more than half of Hawaii's privately owned land; white directors of the Bishop estate controlled another sixth of this land, and part-Hawaiians, Hawaiians, and Asians another sixth. The majority of Hawaiians lived in new urban slums around Honolulu and Hilo, and their population was rapidly declining. The number of full-blooded Hawaiians declined from an estimated 142,650 in 1826 to approximately 40,000 in 1919, three years before Mead's first visit to the islands.[74]

The Hawaiian government had tried to settle Hawaiians on the soil prior to the HHC Act. When the islands were annexed, the U.S. Congress did not apply its extensive homesteading laws to the territory, since an 1895 Hawaiian Land Act supposedly provided sufficient means of settling small farmers. The act opened up public lands to native Hawaiians, offering 999-year leases. Many Hawaiians, however, could not meet the lease restrictions. Others sold their land to non-Hawaiians. While government land had been subdivided into small blocks of five to fifty acres and sold on a lottery basis, the parcels often fell into the hands of land speculators rather than earnest farmers. The 1900 Hawaii Organic Act further dictated the uses of the public lands "ceded" to the United States, with similar results. Overall, these acts further alienated Hawaiian people from their land.[75]

In the early 1900s, Prince Jonah Kuhio Kalaniana'ole and his supporters sought further ways to revitalize the Hawaiian people. The prince's vision of *aina ho'opulapula* (restoration through the land) resulted in congressional passage of the HHC Act, which President Warren G. Harding signed into law in July 1921. The HHC Act was the first comprehensive measure that addressed the population decline of native Hawaiians. Governor Wallace R. Farrington, chair of the HHC, boasted that "rehabilitation, the Hawaiian homes law, is the open door of opportunity by which the Hawaiians may return to the possession of the land." The principles behind the HHC Act resembled articulations of other settlement projects around the world, in which attempts were made to remold both the landscape and a population's economic and social life. An HHC executive reported in 1922 that if the Hawaiian people "might again be placed upon the lands, where they could till the soil and form the nucleus of an independent citizen farmer class, and become self-supporting and raise healthy, happy families and become home-owners, new blood would be gradually infused into the race."[76]

For its first five experimental years, the HHC Act designated lands on Molokai and Hawaii for the return of native Hawaiians, providing between 20 and 80 acres of agricultural land, 100 and 500 acres of first-class pastoral lands, and 250 and 1,000 acres of second-class pastoral lands for each community. Settlers could lease lands for ninety-nine years with an annual rental for each tract of one dollar per year.[77]

At the end of June 1922, Mead visited the Hawaiian Islands to make general investigations into the rehabilitation project. He was at first pessimistic about the future of the HHC's work. Throughout the islands, the government seemed to exist only to "furnish cheap labor for the sugar planters." Overall, he thought that rural conditions were "nothing but commercial feudalism," with Hawaiians and immigrant groups serving as pawns of the big sugar planters.[78]

After meeting with the HHC in Honolulu, Mead visited Molokai and inspected the homesteads, gardens, and agricultural lands on which a few families had already settled. Molokai, the fifth-largest of the Hawaiian Islands, covers an area of 261 square miles. Called the lonely isle, it rises in the east nearly five thousand feet to a forested, cloud-capped, and rain-drenched peak. The arid, west end of the island rises only to about one thousand feet. The land between the peaks, noted for its rich red soil and sweeping winds, forms a plateau about five hundred feet above sea level. This plateau was to serve as homesteading country for Hawaiian natives. By 1900, Molokai's native population had dwindled to one thousand, a number that remained stable. Nonetheless, the HHC hoped to repopulate and carve from the island a group of self-laboring Hawaiian cultivators.[79]

The rehabilitation process started in 1922 with the formation the Kalanianaole settlement on Molokai. The HHC chose homesteaders on the basis of at least 50 percent blood lineage and character. Many came directly from Honolulu, where they had been laborers, engineers, mechanics, foremen, clerks, firemen, mail carriers, and students. In 1922, the HHC developed twenty-two farm lots, thirty-three residential lots, and twenty-five hundred acres of community pasture and grazing land. HHC experts completed road surveys, piped in water for irrigation, established a demonstration farm, and started to clear the land.[80]

The Kalanianaole settlement, like California's Delhi and Durham, grew from modest beginnings. "No attempt is made," an HHC report announced, "to provide a small sugar plantation, or hand over a tract to be cultivated under the direction of a head luna while the homesteader sits on his lanai and smokes and sleeps." Instead, as in Mead's California settlements, the purpose of the first Molokai settlement was to "HELP the home builder HELP HIMSELF."[81] Since the government had not yet completed the colony's infrastructure, pioneer homesteaders moved to Kalanianaole under extreme adversity and despite Cooke's protests.[82]

HHC's first Molokai settlement adhered to methods developed by supporters of the Back to the Land movement around the world. Like California and Australia's land settlement acts, the HHC Act attempted "to create a healthful and attractive life for people menaced by the stress of labor competition in cities and unorganised rural life in the country."[83] The HHC, Mead agreed, was venturing onto familiar ground. Settlers' initial hardships would decrease with time and experience, and the benefits to society as a whole outweighed early settlers' sacrifices.[84] Of course, a racial element was involved: some Hawaiian officials justified the allocation of second-class land to the HHC on the grounds that had first-class land been given, "the main object of the [HHC Act] would be defeated as the Hawaiians would not work the land themselves but would have the work done by Japanese."[85]

Mead, who was virulently anti-Asian, tied this idea of self-labor to citizenship and nationality. "It is realized by practically every civilized country of the world today that proper settling of its people upon the land and rooting patriotism in the soil is the political and social problem that is most vital to the welfare of a nation," Mead reported in the *Honolulu Star-Bulletin*.[86] The HHC implicitly advertised homesteading as a national imperative that implied the fulfillment of certain social, political, and even racial ideals.

"Rooting patriotism in the soil" meant planting the American flag on Molokai and replacing traditional governments and cultures with American ways. Annexation in 1898 made American imperialism a fait accompli, but Americans did not seek forcibly to displace native Hawaiians, who were viewed as a "neo-Hawaiian-American race."[87] They inhabited a racial category that could, with some intermixing and improvement, conform to the American one. Their population had dwindled to such low numbers that assimilation would occur rapidly. In practice if not totally in theory, then, Hawaiians stood in contrast to California's Indians and South Africa's black Africans, for whom the progress of the Anglo-Saxon race required subjugation and ultimately removal.

"Rooting patriotism in the soil" also did not mean leaving Hawaiians to their own devices. In its most popular iteration, Hawaiian resettlement could be used to mold a new class of citizen-farmers that would offset the influence of large numbers of Asian landholders.[88] But Molokai's land settlements and this "patriotism" would not evolve naturally, Mead warned. Only the HHC would bring to Hawaiians' attention the "true relation of land to the national life."[89] From "a political standpoint," reported Newell, "the importance of keeping these islands thoroughly American in racial and social affinities can hardly be overestimated." Logically, therefore, the federal government bore responsibility for developing these settlements along American lines.[90]

Just as William Hammond Hall's policies in South Africa entailed transforming nomadic Boer farmers into American-style capitalists, so, too, did rehabilitating native Hawaiians mean imposing on them American values. Mead believed that settling Hawaiians meant stripping them of the idea that a homestead "is only for those who desire to sit under their own Papai and grow their own Taro." Altering home seekers' mind-sets would also discourage the pernicious land speculation that had created a class of migratory farm workers.[91] Despite his earlier rhetoric, Mead's ideal of the taro-cultivating Hawaiian quickly gave way to plans for commercial farming, where crops would enter the Honolulu market.[92]

Working on the tails of American Progressive reform, Mead advocated sound business and agricultural practices for the Molokai settlements. First, as in his Delhi and Durham colonies, he wanted agricultural experts to run the settlements. These men would transform homesteaders—previously wage earners living in crowded tenements in Honolulu and Hilo or plantation workers—into skilled cultivators.[93] One important difference, however, distinguished Hawaiian settlers from most of their Anglo-Saxon counterparts who settled in agricultural colonies around the world between the 1880s and the 1920s. Although the HHC might create a new breed of Hawaiians farmers, it could not make them owners of the land they inhabited.

Second, Mead sought to adhere to the basic principle that had guided his closer settlement policies in Australia: limiting the size of farms and homesteads. He had seen the hard lessons of farmers who lived on large, costly plots of land in California and Australia. "The country or town worker," he noted, "who takes a one or two acre block knows and feels that he is getting a home. He does not have to assume worries of management which a farm imposes." Mead, who advocated garden homes and patches of land for workers even in the cities, hoped that the HHC would "repeat the world's experience in creating small holdings." The irrigated farms in the Palaau tract on Molokai, for example, matched the size of the Delhi settlement's poultry farms. Individuals could increase their holdings as they gained more experience.[94]

Third, Mead warned that one place could not be fully applied as a model for the next. The Durham and Delhi settlements did not provide a perfect model for the HHC, and the HHC's Molokai experiments could not be replicated exactly throughout the rest of the islands. Molokai's good water supplies, for example, made the region well suited for a homestead colony but made the island "no guide" to future settlements. Access to transportation and markets similarly varied.[95]

Mead also visited some of the HHC's projects on the island of Hawaii. The Waiakea homesteads disappointed him. While a good road built by the territorial government passed through the land, nearly every field showed neglect. In some cases, Mead reported, the cane was "almost smothered with grass." He concluded that a single individual would cultivate a maximum of ten acres. The residence lots, by contrast, contained neat, attractive gardens. The HHC owned six thousand more acres of land adjoining the lots that could also be developed into small workers' homes.[96] "Inquiries showed," Mead wrote, "that men who were subject to temptation, becoming dissipated in town, were leading orderly, sober lives."[97] But his trip revealed mixed results. On a visit to a dairy farm, Mead praised the beautiful herd of cattle and modern cream separators, capping machines, and churns. But when he interviewed Auhana, a Chinese-Hawaiian who resided on the same land, Mead realized that irrigated cultivation had halted completely.[98]

Despite such inconsistencies, Mead felt that the HHC's project could succeed if certain conditions were met. The settlements needed small, irrigated tracts, hardworking settlers, government support, good transportation and markets, and most important, a capitalist mentality. After Mead's optimistic assessment of the Molokai settlements, Farrington concluded that Mead's visit had been "of great value to the Territory."[99]

Yet Mead's confidence about the HHC's future was premature. In early 1923, the HHC again called on the Californian. The settlements had not progressed as planned, and the HHC sought Mead's advice. In April, Mead returned to Hawaii to examine the progress of the Kalanianaole and Palaau settlements on Molokai, identify solutions to the problems that plagued them, and help design future settlements.

The problems that the HHC homesteads faced resembled those in other newly developing regions — technology, markets, transportation, and labor. Irrigation on Kalanianaole presented the first difficulty. Water from the mountains was piped down to points of storage and distribution. Artesian and well water was also pumped from beneath the ground without draw-ing up the salt water. Yet the piping could not carry nearly enough water to homes because of mineral buildup in the pipes. Similarly, the region's fertile soil supported alfalfa, potatoes, watermelons, corn, and tomatoes as well as cattle, hogs, and poultry. The unreliable water supply and prevalence of pests and diseases, however, undermined even the best-planned agricul-tural enterprises.[100]

As in Australia and South Africa, markets also created a dilemma for would-be farmers. Although the Molokai homesteaders could enter the Honolulu market, small Asian homesteaders and truck farming glutted other potential markets. Small markets at Kaunakakai, local ranches, and plantation camps provided few outlets and made systematic marketing difficult. Complained one HHC employee, "Our inability to anticipate the yield, found out by bitter experience, is the key to our marketing troubles. Where homesteaders have quit producing, it has not been their lack of interest, but rather the uncertainty of getting a return."[101] Transportation from homesteads to these local markets was adequate. A decent road to the Kaunakakai pier existed, and several small freighters left for Honolulu each week, although roads within the homestead areas left much to be desired.

The lack of reliable markets only supported the faulty assumption that working small plots of land could transform Hawaiians into small capi-talist farmers. Like the management of Mead's Delhi and Durham settle-ments, the HHC wished to create conditions of independence for the homesteaders. Yet outside of building the infrastructure for settlement, the HHC's role in supporting the settlers and marketing the crops remained vague. Critics commonly charged that things were done *for* rather than *with* the homesteaders. The eventual exhaustion of the home loan fund, as well as the matter of inheritance and indeterminate position of future genera-tions of Hawaiians, also called the HHC's role into question.[102]

Most troubling of all was the homesteaders' growing reliance on Molo-kai's pineapple companies. Many Hawaiians who could not make ends meet leased their holdings to pineapple growers, including Libby, McNeill, and Libby. The HHC's annual reports through 1927 show the homesteaders "working hard and enthusiastically." But after the pineapple companies started renting the homesteaders' land and hiring the homesteaders for wages, the ideal of self-labor deteriorated. Other Hawaiian homesteaders employed Filipinos, Koreans, and Japanese to work in the fields.[103]

These problems arose as the Hawaiian subsistence economy transitioned into a commercial one. One Hawaiian mourned that the islands' "older sys-tem of cooperative work broke down almost entirely. . . . A commercial attitude grew up among the families, to the great regret of many of the people." One settler more sympathetic to the idea of earning a profit asked, "Why should we work, when we don't see the money?—it goes into other people's pockets." To the HHC, this attitude suggested a deep-seated antipathy toward work. Hawaiians, like South Africa's Dutch Boer population, had historically been small cultivators with a correspondingly small desire for profit. One homesteader summed up the general sentiment by asking, "What do you think I'd look like walking behind a plow?"[104]

The issue of self-labor related to the thorny problem of race. By 1905, more than 44,000 Chinese and 111,000 Japanese had immigrated to Hawaii to work in the cane fields.[105] Mead scorned the "alien civilization . . . who will have no permanent interest in the land."[106] Just as Chinese and Japanese laborers competed with whites for work in California (and thus inspired Mead's frequent venomous attacks against foreign labor), Asian immigrants competed with Hawaiians, pushing them into menial positions or unemployment.

Rehabilitating Hawaiians on the land presented a feasible solution to this influx of foreign labor. Identifying Hawaiians as Americans, however, raised complicated racial issues. Even though the HHC attempted to transform Hawaiians into American-style farmers, whites were still seen as a far more advanced and superior race. One visitor noted that Hawaiians "still lived in the stone age, knowing no metals, yet governmentally developed along feudal lines to a degree comparable to that of England a thousand years ago."[107] Yet by the early 1900s, Hawaiians *were* Americans, assigning their status in the overall American racial hierarchy to an indeterminate position.

Mead sidestepped the issue of the Hawaiian race by drawing on his nationalist principles: "No race that does not intermarry and mingle freely with the whites, is or can be a good American. It is not a question of political but social and economic domination."[108] Assimilation could solve the Hawaiian race problem. The San Francisco Chamber of Commerce's Craig argued that "many years may elapse before the racial characteristics of the natives have disappeared, and their gradual absorption into the great American nation, under the unrestricted teachings of the most perfect of all civilizations. . . . When that process is continued to completion, under enlightened and Christian influences . . . it carries with it a train of humanizing benefits."[109] Many Hawaiians intermarried with whites and thus became more "white," joining the islands' increasingly powerful white society.[110] Hawaiians also commingled with other groups, particularly the Japanese. "It seems safe to conclude," one U.S. government official claimed, "that the ultimate Hawaiian-American ... will be something near one-third Japanese, one-fifth Filipino, one-ninth Portuguese, one-tenth Hawaiian, one-twelfth Chinese, [and] one-fifteenth Anglo-Saxon."[111] As far as the HHC was concerned, however, such interracial mixing diluted native blood and thus impeded rehabilitation efforts by shrinking the pool of eligible applicants for HHC land.

The Molokai homesteads thus faced many of the obstacles present at Delhi and Durham as well as others relating to Hawaiian annexation, American nationality, and race. Settlers held varying opinions about the homesteading project. Some seemed "satisfied" with general conditions, but others felt "destitute and in a deplorable condition," "baffled and fooled" by the HHC's capitalistic interest and the pineapple companies. Others claimed that the Molokai settlements were simply experiencing growing pains.[112]

Hawaiian settlement nonetheless proceeded. In 1924, the HHC opened more homesteads for settlement in central Molokai. The Hoolehua-Palaau tract alone boasted 153 individual homesteads centered on a church and school and 6,630 acres of community pasture. By 1924, 287 men, women, and children lived on HHC's Molokai lands.[113] Although these settlements revealed the difficulties of putting city dwellers on farmland, Mead concluded in 1927 that the results "have been so encouraging that the practically unanimous vote of the Legislature in favor of continuing this experiment seems justified."[114] By 1929, the HHC had received 925 applications and had issued 378 allotments on the islands of Molokai and Hawaii, a substantial figure given the number of those eligible for land.[115]

Despite Mead's optimism, the Molokai colonies floundered for many years. Poor markets, saline soil, unproductive lands, problems with irrigation technology, and the government's uncertain role in settling farmers put the rehabilitation program on an unstable path. As Hawaiians intermarried, the 50 percent blood requirement also decreased the number of eligible applicants. Yet the HHC continued its work, eventually folding into the Department of Hawaiian Home Lands under the jurisdiction of the U.S. government.

Hawaii, like British India, California, Australia, and South Africa, joined the world economy in ways that confirmed many of the ideals of progress, including the application of science to agriculture and commercial development. Yet once again, technical and capitalist growth took precedence over social equality, and racial motives contributed to both the sugar industry and irrigated homesteads' evolution. From Spreckels's Maui plantation to Mead's involvement with the HHC, the history of the Hawaiian Islands in the late nineteenth and early twentieth century reveals how commercial development generated new racial hierarchies and inequities. The modern yeoman ideal that Californians persuasively if not always successfully exported around the world had little relevance to Hawaii's agricultural development except when it came to the issue of Americanness.

Not all communities in regions settled by Europeans, however, wished to develop their countries along capitalist lines. Across the world, another people solicited the visionary Mead's aid, this time for Zionist settlement in the fabled land of milk and honey, Palestine.

### CHAPTER 6

# Palestine's Peculiar Social Experiments

In 1923, Elwood Mead left his teaching position at the University of California at Berkeley and embarked on a trip around the world. Since his work in Australia, foreign government officials had solicited his advice on irrigation and settlement policy. After visiting Hawaii and recommending that the government resettle native Hawaiians on small, intensively cultivated farms, Mead sailed to Sydney. He spent four months in Australia, where he helped the New South Wales government develop lands watered by the Murray-Murrumbidgee River system. From Australia, Mead and his children sailed to Singapore, Java, and Calcutta. They then traveled by rail across India, stopping to examine irrigation works on the Ganges River— dams and canals even more impressive than when George Davidson had visited in 1875. After reaching Bombay, the Meads boarded a ship for the journey through the Suez Canal to Port Said and thence to Palestine. They arrived in Jerusalem in November 1923, more than a year after Palestine had come under British mandate. Zionist authorities in the United States and Palestine had recruited Mead's aid for a religious and ideological endeavor—the building up of a Jewish national homeland. After touring the region, Mead reported to the Zionist Executive in London on Palestine's agricultural possibilities. He subsequently returned to Palestine in 1927 to finish the work he had begun.[1]

Agricultural development in Palestine's Jewish settlements offers an additional example of how California's engineers tried to draw distant regions into a world economy and how Palestine reached beyond its borders for agricultural models. Political, cultural, and economic models rooted in the American as well as Western and Eastern European experience consciously shaped the organization of Jewish Palestine.[2] Agricultural development also illustrates how the idea of progress changed in the interwar years to reflect new nationalist and ideological projects such as Zionism, which theoretically created better conditions for Mead's ideas about rural life than existed anywhere else. Attracting people did not present the problems that it had, for example, in Australia. Jews from Europe and America came in such great numbers that small farms were a necessity. Sporadic Arab hostility made cooperative endeavors among settlers essential. Land speculation was impossible because by the time of Mead's visit, immigrants could acquire land only through various quasi-governmental Zionist organizations.[3]

Despite these conditions, Mead encountered one obstacle: many Jewish pioneers were more interested in fulfilling their nationalist and often socialist ideals than in the business of growing crops. Settlers wished to organize rural life along collectivist principles in the belief that only a classless society of agricultural workers could build a Jewish homeland.[4] Mead, however, viewed rural development as primarily a capitalist enterprise. His experiences in Palestine thus stood in contrast to his work in California, Australia, and Hawaii, which took place against the backdrop of capitalist development—the motive at the heart of nineteenth-century progress. The general absence of capitalist and technical growth in the Jewish socialist colonies reveals the importance of these capitalist ideas to the global spread of progress.

### Palestine's Socialist Experiments

The general process that late-nineteenth- and early-twentieth-century Jewish immigrants to Palestine faced resembled aspects of the white settler experience in California, Australia, South Africa, and Hawaii. Jews wished to settle and cultivate a semiarid, fertile land, replace existing ways of government and culture with their own, and achieve a balance between the life and industry of the cities and the country.

Yet the Jewish *aliyah* (movement to Palestine) was unique. Settlers hoped to combine a new nationalism with the traditional idea of a return to Zion to foster the revival of their people on their land on an agricultural basis.[5] But the socialist and individualist colonies that formed between the 1880s and 1920s, from the Bilu to the Zionist settlements, remained somewhat colonial, with white Europeans living among and employing a relatively poor Arab population. Most settlers could not survive without heavy subsidies from Jewish agencies, which undercut the settlers' self-laboring ideals. To exacerbate matters, the socialist philosophy prevalent in many of the settlements often blinded colonists to the necessities of modern machinery, scientific expertise, expansive transportation systems and markets, and above all a capitalist mentality.

Idealism had characterized much of Jewish settlement for more than three thousand years. In the sixteenth century, after their expulsion from Spain, Jews trickled into Palestine. They started to reclaim Palestine as their homeland in the late nineteenth century. The spread of anti-Semitism in Europe, marked by the Dreyfus case and the Russian pogroms after the 1882 assassination of Czar Alexander II, spurred Jewish immigration to the United States and Palestine. In the late 1880s, more than twenty-five thousand Jews from Eastern Europe immigrated to Palestine, governed by the Ottoman Empire from its centers of Beirut, Jerusalem, and Damascus. By the outbreak of World War I in 1914, between sixty thousand and eighty-five thousand Jews lived in Palestine.[6]

The first wave of Jewish settlers in the 1880s formed part of the Bilu movement, a group composed mainly of Russian university students who drew on the ideas of radical Russian thinkers of the time. Writers, including Leo Tolstoy, praised the simple, moral life of the Russian peasant. The Narodniki (Populists) followed this line, stressing the need for Jews to reorganize their lives along more natural lines. Influenced by these ideas, Bilu settlers set out to create a Jewish society based on collective rural life. Though not peasants, these settlers nonetheless saw clear links among collective agriculture, Palestine's economic future, and the revival of Jewish culture.[7]

In the early 1880s, Bilu pioneers formed the settlements of Rishon Le-Zion (First of Zion) south of Jaffa, Zikhron Ya'akov in Samaria, Rosh Pinna in the Galilee, Yesud HaMa'ala near Lake Hula, and Gedera in southern Judea. Filled with hope but with very little practical farming experience, the colonists—many of them former professors, lawyers, and merchants—faced new challenges. Just like many of the settlers in California, Australia, South Africa, and Hawaii, the new arrivals to Palestine lacked the tools and experience needed to transform a semiarid landscape into corn and wheat fields like those in Europe.

Settlers saw desert, ancient terraces, and plains dotted with malaria-ridden swamps. Constant grazing had denuded the hills. One colonist described the Valley of Esdraelon as a desolate plain of five "small and squalid Arab villages."[8] Visiting in 1867, Mark Twain described Palestine as "a hopeless, dreary, heartbroken land" over which "broods the spell of a curse that has withered its fields and fettered its energies."[9] In 1923, forty years after the founding of the first Jewish settlements, Mead noted that Palestine, once "thickly settled in its more fertile areas, stands today a poor and relatively undeveloped country . . . impaired by centuries of wasteful [Arab] cultivation."[10] Small-scale cultivation and irrigation had hardly "improved" the land.

The borders of Palestine drawn during the years of British rule (1918–48) included Lebanon to the north; the Jordan River, Dead Sea, and Arava Valley to the east; the Mediterranean Sea and Sinai Peninsula to the west; and the Gulf of Eilat to the south. The country covers 10,162 square miles, an area roughly equivalent to the state of New Jersey. Although parts of the land are fertile, Palestine offers limited horticultural opportunities. From north to south, the central portion of Palestine is a system of steep limestone hills and mountains, interrupted by the Plain of Esdraelon (the Emek Valley) and the Valley of Jezreel. Only the Jordan River waters the alluvial soils of the plains. In many areas, the country has poor drainage; hot, dry winds throughout the year; and a limited water supply. Palestine receives an average of twenty inches of rain annually and has some underground sources of water. It lacks rainfall, however, between June and September. The remains of Roman aqueducts provided relatively small water supplies, but in the first half of the nineteenth century, agriculture by the predominantly Arab population was traditional and involved low levels of technology, with almost no irrigation.[11]

The early Jewish colonists' efforts at agricultural colonization in this harsh environment would have failed completely had it not been for the paternalistic Baron Edmond de Rothschild, head of the Paris branch of the famous banking family. He paid the colonies' debts, acquired more land for Jewish settlements, and sent trained French horticulturists to study the land. These experts substituted viticulture for grain raising and taught the settlers to grow fine French grapes. With an investment of more than fifty million dollars by the late 1890s, the Rothschilds managed twenty-two colonies with a population that surpassed five thousand and spread over seventy-five-thousand acres.[12] Yet such philanthropy robbed the settlers of their socialist philosophy. Instead of producing a subsistence lifestyle like that of the neighboring Arabs, Jewish settlers became highly subsidized small farmers selling luxury commodities such as grapes and wine. The hiring of cheap Arab labor for the vineyards also undermined the ideal of self-labor.[13]

Capitalist enterprise proved a more feasible lifestyle. In 1900, Rothschild transferred the management of the colonies to the Jewish Colonization Association (JCA), an organization founded in 1891. Later, the Palestine Jewish Colonization Association continued the work that Rothschild had started. The JCA immediately abolished subsidies for the settlers and then established an instruction farm in the Upper Galilee and introduced a system that made settlers landowners only after they had obtained practical farming experience.[14]

In the late 1890s, Zionism changed the nature of Jewish settlement. Theodor Herzl, a young man from a prosperous Viennese family, founded the World Zionist Organization in 1897. Although Jews had assimilated into the larger Austro-Hungarian culture, the general population resented their relatively high economic status, and anti-Semitism increased. Addressing the Jewish plight, Herzl wished to link modern social justice to an old communitarian tradition by establishing a national homeland that would foster Jewish culture. Zionism thus provided a cultural rather than purely economic dynamic for settlement. It spurred immigration from middle- and lower-class homes in Eastern Europe and resulted in the settlement of thousands of Jews in socialist and individualist colonies in Palestine.[15]

Zionism contained utopian and practical, capitalist and socialist elements. Like Alfred Deakin in his desire to implement new land and social policies, Herzl cited Henry George. While critics of Zionism dismissed Herzl's ideas as fantasy, its proponents embraced them. Along with other leading Zionists and other supporters of the Back to the Land movement, Herzl claimed that farmers led happier, healthier, and morally superior lives than did city dwellers. By planting roots deep in the soil, Jewish workers could build the foundations of a nation. But Herzl believed that agriculture, whether undertaken by individuals or socialists, would have to be cooperative to succeed.[16]

Purchasing land underpinned the Zionist project. After fruitlessly appealing to European leaders for support and rejecting Britain's offer of a homeland in Uganda, the Zionist Organization received a charter for colonization from the British, much like those granted to the East India Company and British South Africa Company. The Zionist Organization then began its land-purchasing and settlement work. The Zionist Organization became a provisional quasi-governmental agency based in London, with the Jewish National Fund (JNF) as its most important branch. The JNF worked to develop a national land policy in Palestine, build a network of agricultural cooperatives and training farms, and support the Jewish labor movement. In 1905, the JNF started to acquire land from large landowners and prominent Arab families and lease it to Jewish settlers for cultivation. Four years later, it produced the first communal settlement (*kvutzah*), Degania, on the west bank of the Jordan River, south of Tiberias.

The *kvutzah* (which evolved into the kibbutz) differed from the more prevalent way of dividing land in settler societies. Some enterprises, such as the Chaffey brothers' irrigation colonies in California and Australia, were based on common endeavor but private landownership. Throughout most of California, the government had simply parceled out land to individual owners. Other ventures, like the Hawaiian Home Commission's work, articulated capitalist principles set against government landownership. The experimental *kvutzah* met with favor and financial support from even the nonsocialist sectors of the Zionist movement. Most Jewish leaders realized that in the first stages of colonization, collective efforts would produce greater results than individuals alone could generate. Most settlers lacked the agricultural training that rendered them fit to work independently. Group houses, stables, and fields required smaller initial capital investment than did individual settlement. Many middle-class Zionists also wished to avoid class struggle, which was producing devastating effects in Europe. They thus considered the *kvutzah* and the nationalization of land— a much more radical step than the nationalization of waters in British India and Australia—the best means of upholding social ideals.[17]

While many settlements gave only passing nods to global socialist movements, the *kvutzah* adhered to socialist principles. One American visitor marveled that "there is no private property, tenancy of the land. . . . Members thus own nothing, but, being partners in the community, have a right to everything. There is no money in circulation within the settlement. . . . The scavenger and doctor have the same rights. . . . The government and administration of the communal settlements are democratic." It did not matter if some members smoked more, used more soap, or sent more letters. They worked on terms of equality, addressing each other as "comrade" and receiving basic living amenities in lieu of wages.[18]

By 1904, more than fifty thousand Jews resided in both communal and individualistic colonies run by the Palestine Jewish Colonization Association, the JCA, and the Zionist Organization. Agricultural growth accompanied this increase in settlement. Jewish settlers established scientific experiment stations as well as drained and reforested small swampy areas with eucalyptus. They began to cultivate oranges, raise cows, and construct irrigation systems. Settlers built twenty Jewish villages and schools in the Esdraelon Valley. "What five years ago was little better than a wilderness," one colonist remarked, "is being transformed before our eyes into a smiling countryside."[19]

By 1919, four years before Mead's visit, the national colonization patchwork was starting to tear at the seams. Settlement by the various Jewish agencies had not advanced in an orderly manner. Colonists and Zionist authorities lacked consensus about the ideal nature of the settlements: Was the future of the Jewish colonies to be socialist or individualist? No uniform methods existed for attracting Jewish settlers, financing land purchases, or developing agricultural schools. The labor situation also undercut national ideals. Most of the Jewish settlements could not afford to purchase the modern plows and threshers that would make excess hands in labor-intensive orchards unnecessary. Despite their intention of working the land themselves, many Jewish colonists took advantage of cheap Arab labor. By the 1920s, only a few Jewish settlements in the Galilee had achieved the ideal of the self-reliant cultivator.[20]

The Balfour Declaration spurred new efforts at colonization after World War I. Despite the official sanction of a Jewish state and British aid in paving roads, building up the booming city of Tel Aviv, and improving the port at the old Arabic city of Jaffa, attempts at colonization met more obstacles. By 1922, the number of agricultural settlements had grown to seventy-three, but new immigrants flocked to Jerusalem and Tel Aviv. Zionist authorities became concerned about the poverty of the immigrants to rural areas, their lack of farming experience, and the idealism that substituted for practicality. These concerns led officials to bring in outside experts in 1923.[21]

### Elwood Mead Visits Palestine, 1923

In 1923, at the request of the Zionist Organization, Mead visited Palestine. After extensive investigations into the various social, economic, and agricultural aspects of the Jewish settlements, he concluded that the individualist colonies generally fared better than the socialist ones. The poor conditions of the latter only bolstered his belief that agricultural settlements, particularly in their early stages, needed to be guided by capitalist business practices rather than idealistic dreams. He thus recommended that Zionist authorities create further settlements following the private but cooperative model.

New York leaders of the Zionist Organization, concerned with how settlement was proceeding in Palestine, had first established contact with Mead in 1920. They had learned of his expertise through Jewish settlers who had attended the University of California at Berkeley. "These young men," Mead wrote, "had come to this remote institution to study farming and the science of agriculture at the dictates of a spiritual urge. They wished to take a leading and helpful part in the creation of a national home." To do so, he continued, they chose for their training "a country with an agriculture highly developed but with a climate and products like their own."[22]

Dr. Selig E. H. Soskin, an agronomist who directed the first of the Zion-

ist Organization's intensive settlements in the Jezreel, Jordan, and Esdraelon Valleys, argued that the fulfillment of Zionist aspirations depended on
the success of the agricultural settlements and on tried-and-true knowledge. Soskin petitioned U.S. Supreme Court justice Louis Brandeis to per-
suade Mead to come to Palestine as soon as possible. Otto Warburg, a professor of botany at Berlin's Humboldt University and later founder of the
Department of Botany at the Hebrew University of Jerusalem, was serving
as part of a delegation of Zionists on a U.S. fund-raising tour. Warburg met
with Mead in San Francisco in early 1922 and subsequently visited the Delhi
and Durham settlements, where, according to Mead, Warburg would see
"where California is attempting to solve the same problems that confront
Palestine." Warburg and his colleague, Arthur Ruppin, who headed the
Palestine Office of the Zionist Organization, then convinced Mead to consult for the Zionist Organization and visit Palestine.[23]

Mead arrived in Palestine in late 1923 and toured the country. Just as
other engineers had made general analogies they could understand—
William Hammond Hall had compared Cape Colony to Southern California, and George Davidson had likened parts of British India to California's
Central Valley—Mead described Palestine as a "replica of southern California."[24] He noted that the coastal plain from Gaza to Haifa had its Southern California counterpart from San Diego to Santa Barbara. Both regions,
with irrigation, permitted the growth of all kinds of semitropical fruit,
most notably citrus. The Jewish settlements along the coastal plain grew
oranges, almonds, grapes, olives, and eucalyptus and oak trees. The similarities between the two regions extended to their valleys. The soil fertility
in the Esdraelon and Jezreel Valleys resembled that of the Los Angeles and
San Bernardino Basins. Finally, Mead compared Palestine's Jordan Valley to
California's Imperial Valley, since both lay below sea level. Irrigation thus
created similar possibilities for agricultural development.[25]

Mead identified the three major regions in Palestine capable of supporting his vision of intensive, irrigated settlements: the coastal plain; the Emek,
or Plain of Esdraelon; and the Jordan Valley, from the northern limits of the
Hula Marshes to the Dead Sea. Water was the limiting factor. The Jordan
River, equivalent in size to one of the larger San Joaquin Valley streams, such
as the Merced or Stanislaus, was the only perennial waterway that could be
pumped. The Nuris springs in the Emek provided only a limited source of
water. In the coastal plain, the springs above Haifa could be tapped by sink-
ing wells and pumping water. Mead estimated that at most 3 million of the
region's 5.5 million acres could be cultivated. He agreed with one contem-

porary who noted that as in California, Palestine's greatest possibilities "lie in a very intensive cultivation of small, irrigated plots of land."[26]

Mead praised the way that Jewish agricultural settlement had progressed despite these environmental limitations. Working through the JNF, the Zionist Organization had created thirty-nine of the eighty Jewish colonies established by the 1920s. Twenty-three were socialist organizations, covering a total area of 163,163 acres. Six were individualist settlements, or small holders' villages, known as moshavim; among them, Mead praised Nahalal, the most important.

Spreading across two thousand acres, Nahalal boasted eighty twenty-five-acre farms radiating from the colony like spokes of a wheel. Settlers channeled water for irrigation from the springs in the Kishon Valley, and production and marketing remained cooperative endeavors. Nahalal also had a training school for women, who, Mead observed, partnered with men in all farm and household chores. The settlements, one visitor noted, "have probably done more towards the solution of the problem of women than any other movement in the world."[27] Mead also observed that "nowhere will such training have greater value," since women were crucial to the realization of the Zionist dream. However, Mead criticized Nahalal's inefficient and outdated farm machinery.[28]

Mead, who admired the philosophical ideals behind Jewish colonization, predicted long-lasting prosperity for some of the colonies, including Nahalal. Yet his experiences at Delhi and Durham gave him much to criticize in the overall way that Jewish settlement had proceeded. The various Jewish organizations failed to put most of their settlements, socialist and individualist alike, on a solid business basis. The private colonies, however, generally fared much better than the socialist ones. But for settlers in both the *kvutzah* and moshav, the idea of reclaiming their homeland seemed to provide a far more compelling reason to work than any financial incentive. Mead believed that this idealism hindered the creation of a vibrant agricultural economy. Only the British mandate government brought some semblance of economic order to the country through taxation, the maintenance of law and order, and the construction of roads and ports.[29]

Mead's experience building settlements in California and Australia showed that people needed to invest a certain amount of money—between 10 and 25 percent—in their ventures to guarantee their commitment. In Palestine, however, the Zionist Organization retained title to the land, just as the Hawaiian Homes Commission had on Molokai and Hawaii. Because settlers had little money to invest, the colonies had to finance the settlers and then assist them until they could support themselves. Mead advised that the Zionist Organization amend its policies to reflect business and scientific practices, hire competent advisers for each settlement, build more training schools, construct more irrigation and drainage systems, and design business contracts that clearly stated settlers' obligations. Finally, he suggested that the Zionist Organization and the JNF plan future colonies, farms, and settlements, establishing one that would serve as the standard for others.[30]

Despite his general criticisms, Mead praised the idea that inspired Palestine's experiments in collective agriculture, experiments that later gave rise to the modern kibbutz movement. As he prepared to return to California, Mead portrayed an optimistic future for Jewish settlement. "The contrast between one of these Hebrew colonies and the nearby Arabic village," he wrote, "is one of many centuries' duration. In the mud huts and in the fields of the fellaheen, life has stood still. . . . The fields are still scratched with a crooked stick, the grain harvested with sickles and threshed by the trampling feet of goats, as in the time of Boaz. The Jewish villages have been cleverly planned by European experts. . . . The homes are simple and small, but they are modern."[31] When Mead returned to the United States, he publicly endorsed the Zionist program and blamed Ottoman, Islamic, and Arab culture for the demise of the Roman irrigation system that had once made Palestine the land of milk and honey. "It is hard to place a limit," Mead concluded, "on what Jewish sagacity, money, and enthusiasm will do."[32] With systematic settlement, irrigation, and sound business principles, Jewish settlers could "create conditions of life as attractive as can be found in any part of the world."[33]

## The Joint Palestine Survey Commission

Mead's report greatly pleased Chaim Weizmann, president of the World Zionist Organization and later the first president of the state of Israel, who used it as justification for focusing Zionist efforts on Palestine rather than throughout the diaspora. But Mead's confidence in the future of the Jewish colonies was short-lived. The problems that he identified continued to plague the Zionist Organization's settlement plans. In 1926, Weizmann and Mark Schwartz, director of the Palestine Department of the Zionist Organization of America, again solicited Mead's advice. American Zionists in particular had become concerned that too much money was being wasted on social projects rather than self-sustaining economic activities. Prominent American philanthropists, including Lee K. Frankel, president of New

York Life; Louis Marshall, a renowned attorney; and Julius Rosenwald, the former head of Montgomery Ward, had decided to withhold funds for Jewish colonization until Palestine's land settlement policies were properly studied and put on economically sound footing.

The meeting involving Weizmann, Schwartz, and Mead resulted in the creation of the Joint Palestine Survey Commission, charged with identifying the problems associated with agricultural colonization in Palestine. Zionist authorities set up the commission as a non-Jewish, non-Zionist body to avoid biasing its findings. Mead, then serving as the commissioner of reclamation in the U.S. Department of the Interior, solicited the aid of a host of experts: Jacob G. Lipman, director of the Agricultural Experiment Station at the New Jersey College of Agriculture; Arthur T. Strahorn, a soil technologist with the U.S. Department of Agriculture; Frank Adams, Mead's colleague in the Department of Irrigation Investigations and Practice at the University of California; Knowles A. Ryerson, a horticulturist with the Agricultural Experiment Station in Haiti and later dean of agriculture at the University of California; and Cyril Q. Henriques, an irrigation engineer with the Zionist Organization in Palestine and a former irrigation engineer in India. In the summer of 1927, the commission, representing American tenets of scientific management and professionalism, headed to Palestine to investigate the Zionist Organization's complaints.[34]

The commission's final report, which examined economic development, water management, and settlement as interrelated building blocks of a nation, stunned Weizmann. Mead still wished to see Zionist efforts succeed. But in contrast to his earlier report, he recommended against the creation of any new Jewish settlements, particularly noncapitalist ones, and advised that much of the colonization efforts should be written off. Overall, the commission's findings revealed how acutely the Zionists' social beliefs had hindered the growth of commercial agriculture. Philosophical idealism blinded the settlers to their own obstacles. Attracted to a larger idealistic movement, settlements had more of an emotional, philosophical, and religious goal than an economic purpose.[35]

In California, Adams noted, settlements had established and mostly followed well-defined principles of colonization. Embodied in theory by Delhi and Durham, these standards included the careful selection of qualified settlers with at least some agricultural experience and capital, the preparation of the land for agriculture prior to sale, advances to colonists for improvements, the offering of expert advice and education, and the sale of land at low interest rates. Private ownership, in turn, conferred the feeling that settlers were, as Mead noted, "permanent and respected members of the community in which they live," individually responsible for the success of the community. Other factors, of course, contributed to each settlement's relative success or failure, but Mead believed that none could hope to thrive without these sound business practices.[36]

Zionist colonization, by contrast, had progressed haphazardly. Private donations and philanthropy rather than established government agencies sustained the settlements. Few settlements enjoyed financial independence from the Zionist treasury, a body that relied on the whims of its benefactors and lacked previous colonizing experience.[37]

The Joint Palestine Survey Commission explored all aspects of Jewish agricultural settlement in Palestine. Delving into the colonists' training and educational facilities, the experts noted many missteps. Settlers followed inadequate advice from the Colonization Department and planted varieties of fruit trees unsuited to climate and soils.[38] "Because trees behave one way in California or Italy," Ryerson warned, "is no sign that they will behave the same in Palestine."[39] Ryerson recommended that settlers plant crops along Californian lines, such as avocado, citrus, and olives, only where conditions permitted. Horticulture seemed to be coming along well only in one of the valleys, where four graduates of the University of California who had specialized in horticultural studies were aiding settlers' work in the orchards, vineyards, nurseries, and fruit-packing house.[40]

Mead and his colleagues further brought the idea of science and technology to bear on Palestine's agricultural institutions. Lipman proposed that the Colonization Department create a research and education bureaucracy modeled after the U.S. experiment stations. The state and federal support that poured into these bastions of applied research were "essential to the economic and social welfare of the rural communities," Lipman claimed. A college of agriculture, as at the University of California at Berkeley, would coordinate research and teaching and play a significant role in "modernizing . . . agriculture and in determining whether Zionism is to reach its goal without undue groping in the dark."[41] Lipman also recommended improving the Institute of Agriculture and Natural History, a disorganized body headed by Warburg. Both Ryerson and Lipman stressed that all educational and research facilities should be severed from the Colonization Department's political and social agenda.[42]

The commission also examined Palestine's markets, transportation, and agricultural production facilities. The region had three markets: home, neighboring countries (including Syria and Egypt), and distant markets (primarily England). The coastal plain from Haifa to Gaza provided one of the world's finest citrus-growing areas, with the Jaffa orange the most important and valuable of all the fruit grown in Palestine. Arab expertise had already developed the Jaffa orange, with great interest from British exporters, before the Zionist colonization of Palestine. By the time of statehood, Arab-owned and -operated orange groves, about half the total land on which citrus grew, had made the fruit famous. After 1948, many Arab cultivators took over from Jewish growers. But at the time of Mead's visit, England still received most of its orange supply from California, though Palestine had the advantage of closer distance and shorter transit time.[43]

Inadequate transportation hindered Palestine's entry into these profitable markets. The British mandate government paved roads from the Emek to Tel Aviv and Haifa.[44] Yet slow, expensive, and temperamental camels still provided farmers with the chief means of transport to the nearest port or railway station. Ryerson estimated that a one-ton motor truck could do as much work as ten to twelve camels. Furthermore, according to the Commission on Fruit Export, "In the journey from shore to ship's hold the fruit probably suffers worse damage than in the whole of the rest of the four weeks which intervene between the picking of the fruit and their sale in the United Kingdom."[45]

The horticultural industry's lack of modern machinery—to Mead, "part of the march of civilization" and therefore "inevitable in Palestine"—also hampered efficient marketing and production methods.[46] Ryerson pointed out that orchards employed six to eight men where one would suffice. The majority of horticulturists used antiquated Arab plows and camels rather than modern combine harvesters and threshers, hay presses, centrifugal pumps, sprayers, dust guns, tractors, trucks, and automobiles. Only a few of the settlements, such as Petah Tikva, had modern machinery, much of which had come from California.[47]

Growers also showed a disappointing lack of coordination in developing credit facilities for financing and marketing crops. Credit institutions involved either bankruptcy or very high prices to compensate for actual and anticipated losses. Farmers in Palestine, like those in America, also found marketing difficult. A Jewish agricultural cooperative known as Hamishbir, a branch of the socialist Jewish Federation of Labor (JFL), theoretically controlled the marketing efforts of the Zionist colonies, but it had mixed results.[48] Adams noted that the Jaffa orange, particularly when cultivated in a privately run colony, "is of such superior quality as to be able to meet competition from other countries." But the absence of common marketing and production standards among Arab and Jewish producers spelled disaster. Since farmers lacked the capital to work on individual bases, cooperative enterprise was the key to success, but agricultural cooperation had made little headway. The sole Arab cooperative, for example, had not even survived its registration.[49]

Settlers in India, California, Australia, South Africa, and Hawaii experienced many of the same financing, marketing, and transportation issues. In Palestine, however, Mead believed that the most serious obstacle to profitable agriculture was an idealistic philosophy that hindered economic development. Some of California's settlements, such as the Chaffeys' irrigation colonies, heralded principles such as temperance. Others, including the Hawaiian Homes Commission's resettlement plans, advocated ideals including native rehabilitation and "Americanization." Yet these varied, progressive ideals, many of which disappeared as settlements became more profitable or had begun as double-edged swords, did not impede agricultural growth. By contrast, Mead reported, "Theoretical programs [in Palestine] based on desire and hope, without adequate knowledge and recognition of existing limits and difficulties[,] are largely responsible for present unsatisfactory conditions in the Zionist colonies."[50] Mead did not direct his venom at socialism per se. Rather, he criticized how this system affected the solvency of the moshavim, *kvutzot*, and private colonies.

Mead believed that the heart of the problem for the *kvutzot* lay in the dreams of the socialistic, highly centralized JFL, which intended to establish in Palestine a workers' society free from capitalism and the exploitation of labor. The JFL adhered to the idea that Palestine must become a cooperative commonwealth in which the public owned utilities and producers' and consumers' cooperatives managed all industries and businesses.[51]

By the time Mead and his commission visited Palestine in 1927, the JFL boasted more than twenty-five thousand members and provided extensive services. It acted as an employment agency for immigrant *chaluzim* (pioneers). JFL agents prepared potential *chaluzim* for settlement by organizing classes designed to serve the educational needs of workers, including the Hebrew language, the history of Jews, and professional and technical courses. In accordance with its socialist philosophy, the JFL adopted a friendly policy toward Arab workers, helping to organize and eventually admit them into Histadrut (a Jewish labor trade union, or "Federation of Labor") membership. JFL members believed that Jewish labor could never hope to make progress in Palestine if it attempted to keep native labor in the position of a cheap competitor, as labor movements elsewhere did. Jewish settlers' daily exigencies made manual laborers more valuable than mental laborers. Indeed, one visitor noted, "It is not simply the change in their method of earning a living that is important, but the social implications of that change. [Jewish settlers] have broken down those barriers of snobbery which in the Diaspora effectively restricted the Jew from entering the ranks of manual labour."[52]

In Palestine, the need to provide the basic industries with manual laborers and produce a class of agricultural laborers superseded the need to fill middle-class occupations. The same visitor noted that although Jewish settlers could get Arabs to perform basic manual labor, as in many non-socialist colonies, "the position of the Jew would [then] be not unlike that of the white man in Africa or India; they would be playing the accustomed role of white exploiter of black native labour and probably throw a sop to their conscience by trying to believe in the glorious White Man's burden."[53] Most colonists believed that despite the growth in Arabs' citrus groves, Palestine's resident population of 350,000 Arab fellahin and Bedouins was useless in transforming the desert into a garden.

Mead admired the JFL's principles of self-labor and cooperation. Mead's colleague, Sir John Campbell, likewise acknowledged the Jewish homeland's deep debt to the JFL. At the same time, he believed that the JFL's "preponderating influence," which elicited too much blind loyalty to the movement, led to "undesirable results." Despite its aid to new immigrants, it did not carry out its work efficiently, economically, or with any continuity of plan. The JFL also pressured Zionist organizations and private donors to fund additional settlements to alleviate overcrowding. These extra people sought outside employment, usually with Zionist organizations that then established even more unsustainable settlements.[54]

Mead and his colleagues generally believed in individualist colonization modified by cooperative enterprise. They opposed on principle the guild system of collective work and ownership, as found in the *kvutzot*, which had many weaknesses: it decreased individual members' responsibility and incentive for special efforts as well as settlers' desire to treat commonly owned property with sufficient care.[55] Several "leading [Zionist] officials," Campbell reported, were "more concerned with putting into practice their social and political theories and ideas than with the humdrum business of settling Jewish colonists. . . . [P]ower has been, more or less, completely divorced from responsibility."[56] All of Mead's investigators agreed that the JFL must be separated from the agricultural sector and the settlements reorganized to produce profit.[57]

So far, the economic and self-help motive that Mead thought so important to colonization in California and Australia had not come into play in most of the *kvutzot*, where Jewish organizations completely subsidized the settlers. Although Mead had advocated the principles of closer settlement in much of his work, he noted that many of the colonists lived on such small areas of land that they could not possibly make a living. Settlers instead cheerfully accepted their hardships, the simplicity of their lives, and their plain homes.[58] One settler told Adams that he had nothing to do between seven o'clock in the morning and five o'clock at night except discuss social theories and politics. When Adams asked a labor leader if the constant deficits were worrisome, he replied that the greater the need, the greater the amount they could get from Jews around the world.[59]

Mead argued that Palestine's most successful private colonies, the former Rothschild settlements run by the Palestine Jewish Colonization Association, should serve as a model for the JFL's settlements. These thirty colonies, which had acquired more than 125,000 acres of land, operated according to sound business practices and scientific management. They paid no attention to social theories, eschewed philanthropy, and lacked the "peculiar social experiments" that characterized other Zionist colonies.[60]

Mead wrote the joint commission's final report, basing his recommendations on the policies that he had developed in Australia and put into practice in California and Hawaii. He outlined plans for the appropriate size of farms, which varied according to the type of crop and need for irrigation. Most farms would have to exist on a dry farming basis. He advocated the reforestation of the hills as a national duty. Planting trees, a project the JNF subsequently undertook, would help restore the region's sparse water supplies. Mead also recommended that Zionist organizations prepare land for settlement before colonists arrived. Such planning required about two years for mapping and a soil survey, determining water supplies, subdividing holdings into areas of approximately equal productive value, and constructing roads and drainage systems.[61]

Scientific management, modern farming machinery, and long-range business planning would put settlements on a sound business basis. Mead also created a model contract between JNF and the settlers that fixed the terms and conditions of repayment. "Many [settlers] do not know how much has been expended in their behalf," he warned, "and the obligation to repay is too indefinite to be regarded seriously."[62] A loose relationship between the colonizing agency and settlers had removed the usual incentives to industry. Dr. Leo Wolman, a noted American economist, recommended that the JFL set up collective agreements and arbitration machinery, work out pension and insurance plans, and bring in a horde of experts—a tax expert, a tariff expert, an immigration expert to regulate the flow of new arrivals, and a financial expert to organize a centralized banking system.[63] These officials would be chosen for their administration ability and technical competence, not for their philosophical leanings. The commission condemned the creation of further colonies until those that already existed could support themselves. "Until colonization is placed on a business basis," Ryerson argued, settlement should halt completely.[64]

Most important, Mead and his colleagues criticized the labor movement's detrimental influence on the Zionist farm colonies. In one instance, Mead praised a settler who had shown cleverness in the choice of crops and in the extra prices he had obtained through better marketing methods. Zionist authorities, however, condemned such acts. By trying to get ahead of his neighbors, the settler had exhibited the wrong attitude. Although Mead admired the settlers' sincerity, he regarded their socialist aims as injurious to the development of a region that lacked established markets and commerce. "Settlement," he concluded, "ought to be entirely divorced from any control that would restrict the attractiveness of its life and the breadth of its conception."[65]

The commission's report, which attacked the collective settlements, incurred the wrath of the Left but garnered support from Weizmann and the Right. Debate over Mead's report occupied the Zionist Executive's 1928 Berlin meeting, but a compromise ultimately was reached. The Zionist Executive adopted an amended proposal that recognized Palestine's unique economic situation, legitimized labor's ideology, and strengthened ties between the two camps. Most important, the commission paved the way for the establishment of the official Jewish Agency in 1929. This agency, authorized by the 1922 League of Nations mandate, became the operative arm of the World Zionist Organization and a way to unite Zionists and non-Zionists. (In the 1930s, under future Israeli prime minister David Ben-Gurion, the Zionist Executive and Jewish Agency together formed the backbone of what would become the state of Israel.)

While it is difficult to quantify Mead's exact legacy, his influence extended from agricultural settlement policy and water law to nation-building. His work, which envisioned the Jordan Valley as part of a large-scale economic development program, informed the ideas of Walter Clay Lowdermilk, chief of the Soil Conservation Service of the U.S. Department of Agriculture. In the late 1930s, he proposed a Jordan Valley irrigation project based on the Tennessee Valley Authority's work. Lowdermilk's subsequent book became an important political tool for debate over the economic viability of a future Jewish state. Mead thus helped to open communication lines between Zionists and the U.S. Bureau of Reclamation, Department of Agriculture, and Tennessee Valley Authority, which provided Zionists with key technical information and political relations with the U.S. government.[66]

In the next decade, other Americans tweaked Lowdermilk's plan and worked with the World Zionist Organization to implement irrigation and hydropower schemes. Through the 1950s, Israel drained the Hula Basin in the Upper Jordan Valley, alleviating malarial breeding grounds but creating harmful leaching into the Sea of Galilee. In 1964, Israel started to pump water from the Jordan River to the country's coastal and southern regions, thereby creating a more integrated water network. Mead's warnings about overpumping from the Sea of Galilee and Lower Jordan as well as his warning about the Negev being too arid for cultivation rang true as well.[67]

Mead also influenced water law in Palestine. In the 1940s, the Wyoming water statutes he had developed in the 1890s influenced the British mandate government's policy on surface water usage, including the doctrine of beneficial use, prior appropriation, and rights tied to landownership. Though Mead suggested a variety of models for water management, from irrigation districts to cooperative associations to private water companies, he stressed the need to centralize water authority. Israel's Water Law of 1959, which declared all waters public property, drew directly on his ideas.[68]

Like many of his colleagues and clients, Mead was a visionary who wanted modern rural life to provide an antidote to industrial society. At the same time, he was a shrewd engineer who knew that a region's economic growth required modern technology and transportation, scientific and business planning, and a capitalist mentality on the part of both planners and settlers. Although he was not always right, Mead realized that Jewish settlers would have to sacrifice many of their social ideals to develop profitable commercial agriculture and enter the modern world system.

For the next half century, ideological principles matched by solid business ones helped ensure the success of agricultural and philosophical enterprises such as the kibbutz. Mead did not live to witness the abbreviated life of the modern kibbutz movement. Its demise—a surrender to capitalist exigencies—attests to the defining theme of nineteenth- and twentieth-century progress and its core struggle to reconcile social and technical goals.

# The Common World Destiny

The dream of progress shared by Elwood Mead and other California engineers had deep roots in the nineteenth century and persisted into the first half of the twentieth. This was, after all, the time that Frederick H. Newell dubbed the Age of the Engineer, the era when the idea of universal progress took on different variations around the world.[1] This vision of progress, based largely on technical innovation and capitalist growth, paid little heed to questions of race, nationality, culture, and history. Instead, engineers such as George Morison believed that progress—in the form of such awe-inspiring creations as railroads, steamships, and telegraph cables—would bring people of all races into contact, break down national divisions, and "finally make the human race a single great whole working intelligently in ways and for ends which we cannot yet understand." Technology could surmount the environmental and racial barriers that had curtailed the progress of "backward" countries. Electric trolleys, deep-level mines, and irrigation systems would sweep away centuries-old traditions. As California's engineers circulated the globe with technology, models of development, and ideals in tow, Morison concluded that "the new epoch has barely begun."[2] Technology would, in effect, create what Eric Wolf describes as "common destinies."[3]

Morison was not alone in his thinking. From the laying of the first transatlantic cable, Anglo-European engineers, scientists, businessmen, policy makers, and colonial settlers around the world envisioned that technology, free markets, rule of law, scientific education, and other advancements would put the tribes of Africa on equal footing with the "higher" civilizations of Europe. San Francisco celebrated precisely this idealistic notion of progress, of a world brought together by technology (and Anglo-European conquest), with the 1915 World's Fair.

We now have the historical perspective to understand that this vision was flawed in ways that most of its advocates failed to acknowledge at the time. Nonetheless, it is easy to see why Morison thought the way he did. He lived during an era of economic globalization, laissez-faire, and imperialism. He witnessed Queen Victoria's empire acquire a quarter of the earth's land-mass. He saw Britain's foreign economic interests, in the words of English cotton industrialist Richard Cobden, try to act "on the moral world as the principle of gravitation in the universe—drawing men together, thrusting aside the antagonism of race, and creed, and language."[4] And he watched the United States for the first time secure territorial possessions beyond its continental borders. The speed at which such expansion occurred sug-gested that nations functioned as interrelated units and that a common world destiny, if cast in Anglo-European terms, was indeed possible.[5]

In the late nineteenth century, scientific knowledge and technical exper-tise allowed Europeans to exploit regions that had previously remained iso-lated from Western developments and to confer authority in new ways.[6] Di-verse products, processes, and ideas aided in imperial conquests. Europeans entered the non-Western world armed with vaccines, ships, electricity, and printing presses. Quinine protected European settlers from malaria. Quick-firing breechloaders replaced muzzle-loaders among the forces stationed on the imperial frontiers. And the compound engine, Suez Canal, and sub-marine cable decreased travel and communication times across the oceans. These inventions reduced the time and cost of exploiting new territories. By the 1880s, people who set out to conquer new lands exerted far more power over native peoples and foreign environments than had been the case only twenty years earlier.[7]

This global economy certainly presented unparalleled opportunities for technocrats and ideologues. In particular, California engineers, working in a climate of post-gold-rush technological change, cast their expertise far and wide. William Hammond Hall advised Victoria's Deakin government on irrigation policy. George and Ben Chaffey tried to replicate their South-ern California colony of Ontario in Australia, complete with temperance and scientific education. John Hays Hammond went to South Africa with American ideals about good government, civic independence, and free markets. And Mead applied his agricultural business principles to home-steads in Hawaii and Jewish Palestine.

Why few of the California engineers' exports resulted in a foreign re-gion's wholesale adoption of technology or politics reveals the vagaries of history. Variations in institutional arrangements highlighted the differences between host countries and California. Land tenure, labor systems, political practices, race relations, extent of markets and of transportation systems, and contingencies such as economic depression and war created divergent histories. These factors, combined in different ways, determined the applicability of Californians' exported technologies and their frameworks. A common political culture, for example, facilitated the back-and-forth relationship that characterized the exchanges between California engineers and Australians, unusual in their degree of reciprocity. But in other regions, engineers' transfers posed a more unilateral arrangement than did the more mutual import-export trade among the United States, England, and larger Atlantic world. In South Africa and Palestine, California engineers entered a region where European settlers had modified or replaced legal, economic, and cultural institutions. Such places often lacked an infrastructure receptive to major technological exports. In 1921, hindsight allowed Hammond to sum up this inequality: "The future of the world's commerce depends upon the development of the resources of the backward nations, and the capital required must come, in the near future, at least, chiefly from America."[8]

If the nineteenth century in theory brought places closer, it also cast them farther apart. The Age of Imperialism, which carved the world into "developed" and "less developed" parts, generated the inequalities that fueled a modern economy based in large part on the exploitation of natural resources. Such divisions of land and power had long histories determined by peasant-lord agrarian relations, agricultural and industrial development, and transnational relations. Imperial wars and territorial expansion produced different state regimes and geopolitical hierarchies. State and class relations shaped countries' relative positions. All of these factors put regions on different paths of modernization. As a result, countries did not have or receive equal information and goods. Nor did they apply technical and scientific knowledge in the same ways.[9]

Even if national, social, racial, and cultural differences had been malleable, nature always compliant to human will, technology universally desirable and available, and politics trivial, California engineers still might not have fulfilled their ideals of progress. This state of affairs had much to do with engineers' position, character, and ideology between the 1870s and 1920s, factors that cannot be overlooked when evaluating the relative success and failure of Californians' mining and irrigation projects around the world.

Engineers' professional backgrounds and relative social positions affected the outcomes of their projects. Like any other highly specialized profession, hydraulic, irrigation, and mining engineering drew on a certain type of personality with a specific skill set. In 1921, Hammond sketched a picture of the "classic" engineer in his monograph, *The Engineer*. He was a man "by temperament vigorous, adventurous, and aggressive, . . . who loves a roving life in quest of new things, whose instincts are curious, inventive, creative."[10] He was a member of professional organizations, such as the American Society of Civil Engineers and the American Institute of Mining Engineers. Such societies came of age during the 1870s, when each distinct branch of engineering defined its territory. The engineer read trade journals; *Irrigation Age* and the *Transactions of the American Society of Mining Engineers* offered a wellspring of knowledge, controlled the publication of technical papers, and censored divergent opinions.[11] As the number of professional schools increased, the engineer became increasingly educated. He attended the Colorado School of Mines; the Royal Mining Academy in Freiberg, Germany; or one of the professional programs at Stanford or other such universities. Because he often worked internationally, the engineer learned both English and foreign languages. He also took classes in economics, business, finance, government, law, history, sociology, philosophy, fine arts, and physiology, including personal hygiene, for who knew what conditions might await?

These skills were necessary, Hammond explained, because the engineer dealt with people from all walks of life. He was "necessarily in co-operation with the business man—the capitalist, the promoter, the manufacturer, the merchant, the industrial and commercial man," as Hammond's collaboration with South Africa's mining magnates showed. Nevertheless, he remained "close to the laborer at every turn."[12] Hammond believed that the engineer's logical thinking suited him well for the role of social leader and class arbiter. But like the laborer he reputedly supported in his dealings with capitalists, the engineer should not expect to become wealthy. Engineering was, Hammond lamented, "a profession which a man enters more from love of the work than a desire to get rich," though "he may eventually come to have an interest or a partnership in the business . . . he may become a promoter, and eventually, perhaps, a capitalist himself."[13] Ironically, few engineers got rich faster than Hammond did.

Hammond took great pains to avoid being so blunt, but the engineer often found himself a classic middleman caught between the demands of labor and capital, science and business. Thorstein Veblen, who published *The Engineers and the Price System* in 1921, similarly viewed the engineer as occupying this middle position. Yet while Hammond exalted the engineer's possibilities, Veblen viewed this "technological man" as merely "an awestruck lieutenant of the captain of finance."[14] Indeed, circumstance often pitted the engineer against his employer or financier, as when Hall failed to cooperate with South Africa's Gold Fields Consolidated and the Chaffey brothers struggled to obtain government funding in Victoria. Veblen predicted that engineers would be efficient in their work only if they unshackled themselves from the "price system," organized into their own class, and revolted against the existing social order. Writer and critic Lewis Mumford agreed, noting that the new class of engineers "will, without doubt, constitute the direct and necessary instrument of coalition between men of science and industrialists by which alone the new social order can commence."[15] A cadre of organized technocrats never rose up to challenge the capitalist employers, though in certain circumstances, engineers rebelled independently. But as Hammond's part in the Jameson Raid attests, rebellion often occurred in support of rather than against capitalist principles.

If the engineer was not a social revolutionary in the conventional Marxian sense, he could still, according to Hammond, play an important role in politics. The engineer viewed government "as a vast engineering undertaking, in which the fitness and integrity of men are fundamental necessities," and Hammond predicted that engineers would "bring a broader mind to bear on public problems than will the lawyer, financier, or merchant."[16] To a certain extent, he was right. Hall, whose position as California state engineer acknowledged the importance of centralized planning in a frontier society, raised important social questions about land, water, and settlement patterns. But as a classic middleman, he lacked the clout needed to enact lasting change. And Mead, though he eventually headed the U.S. Reclamation Service, lost support for his state settlement programs in California.

Hammond's sketch of the engineer reflected romantic hopes and dreams for the profession but perhaps little else. Only a handful of his colleagues enjoyed political influence, social standing, and wealth. Even Hammond, one of the century's most esteemed mining engineers and friend and consultant to American presidents and statesmen, left South Africa with his tail between his legs. Still, if engineers were often blamed for problems caused by local politics, labor relations, or the natural environment, they nonetheless earned worldwide respect for their work. Their contributions to society cannot be measured by wealth and status alone.

One notable exception to the engineer's frequently compromised position was the best-known engineer of his day, Herbert Hoover. In the twentieth century, he turned this general respect for the vocation into real political prestige. Like his contemporaries in the 1890s and 1900s, Hoover represented the tenets of professionalism and Progressivism and applied a social approach to the problems of engineering.[17] Trained in geology at

Stanford University, he took a job in mining engineering after graduating in 1895. He pushed ore carts at a gold mine near Nevada City until he joined a British mining firm that sought skilled American mining engineers to work in western Australia. In May 1897, Hoover arrived in Australia to survey mines in the Coolgardie gold fields. A few years later, he departed for China, where he transferred the knowledge he had brought from the United States and applied in Australia to China's Bureau of Mines. High salaries and profitable Burmese silver mines made Hoover a wealthy man. By 1908, he owned a mining consulting business and had published the definitive text on mining engineering. Six years later, he was a self-made millionaire several times over.

Like many of his contemporaries, Hoover dealt with the great social and economic questions of the day. Trapped along with hundreds of other foreigners in Tianjin, China, during the Boxer Rebellion, Hoover joined the city's firefighters. This job led to his leadership role in directing food and water to anti-Boxer Chinese who had taken refuge in the compound, a portent of his later war relief work. When World War I started, Hoover organized the Committee for the Relief of Belgium, raising more than a billion dollars for food and medicine. Under President Woodrow Wilson, Hoover directed the U.S. Food Administration. In 1928, he was elected president of the United States.[18]

By the time that Hoover became president, engineers' roles had evolved considerably since their inauguration as "professional" men in the 1870s. During that half century, engineers became more unified professionally. They introduced and reflected society's larger ideals, such as scientific management and centralized planning, and served as harbingers of major movements, including Progressivism. All engineers combined technical, social, political, and ideological goals, stressing some aspects more than others in their work.

The engineer of the 1870s was an idealist who applied scientific and technical principles to the problems at hand. Although he concerned himself with the major social, political, and economic matters of the day—land monopoly, settlement patterns, agriculture, water rights laws—he was more a scientist. George Davidson, who so greatly admired India's grand hydraulic engineering works, failed to understand their limited political context and relevance to California. Similarly, Hall, though he wrote many laws designed to untangle California's antiquated water laws, applied primarily scientific and technical solutions to the vexing problems of flood, drought, and irrigation. Politicians considered his multivolume tome on water law too academic, too idealistic, and too impractical for California's immediate problems. In this era, idealism trumped feasible technical solutions.

By the 1890s, technology had become much more sophisticated. Engineers started to wield technology as a primary tool to effect social change and to equate such tools with progress. The Chaffey brothers, for example, believed that their extensive irrigation systems in Ontario and Mildura would produce an egalitarian agrarian community; Hammond naively hoped that deep-level mining would bring free markets and civic independence. Engineers wholeheartedly believed—or at least acted as if they believed—that technology could solve society's ills.

In the early decades of the twentieth century, the engineer started to mix hard-nosed business principles and managerial "objectivity" with the same romantic idealism that had characterized earlier decades.[19] By the 1910s, engineers even professed an ideology that paralleled—and sometimes led—the thinking of America's Progressive reformers. Mead, for example, spearheaded major bureaucratic changes in Victoria that inspired similar transformations in America. Still, many engineers shared the popular nostalgia for a simpler, individualistic, and rural social order.[20] They paradoxically instilled this belief with modern business, scientific, management, conservation, and engineering principles that required even more complex technology and bureaucracies.

After World War I, many engineering projects relating to rural development took on strong ideological agendas, which shifted engineers' thinking away from their embrace of openly ideological solutions. The idealistic Mead became more business-minded when he addressed agricultural development in Palestine and tried to rectify some of the Zionist movement's less pragmatic plans. As his application of business principles to agricultural settlements in California, Australia, Hawaii, and Palestine shows, engineering had become, as Hammond predicted, an integral part of business—and vice versa.[21]

Despite the subtle shifts in outlook between the 1870s and 1920s, most engineers maintained their strong moralistic and idealistic assumptions about the world. In many cases, they simply assumed that the laws governing the realm of human affairs—from markets to racial hierarchies— mirrored the scientific laws governing their daily work. This ideology resulted as much from their times as from their individual thinking. Engineers identified their work with the laws of classical economics, with what Edwin Layton calls a "bastard offshoot" of Social Darwinism, and by the early twentieth century, with Taylorism and scientific management.[22] These ideologies reinforced the assumption that scientific solutions could be applied to social problems. Thus, engineering could be thought—just as Hammond argued in *The Engineer*—to include economics, politics, and social change as part of its purview.

Engineers also believed that one of the most important parts of their profession dealt with the nature of the method rather than the specific subject matter. Thus, if most engineers did not understand and could not explain the exact nature of the scientific laws that governed society, most were able to communicate general principles, such as the need to preserve the human race.[23] If people were malleable, often unpredictable beings, they still abided by a set of laws, vague though they were. Newell believed that engineers should examine and treat human groups as if they were machines "in which the wheels and bearings are men and not metals."[24] This reasoning provided strong justification for engineers' ideas about racial hierarchies. In some cases, driving ideas about race even resulted in new institutional arrangements. In Hawaii, for example, the debate over what constituted a Hawaiian resulted in the creation of new agrarian structures designed to put white settlers firmly in control. Similarly, irrigation and mining enterprises in South Africa reflected white settlers', engineers', and administrators' views toward black Africans. These human groups, which did not operate according to engineers' understanding of social organisms, fell prey to the laws of Social Darwinism.

Engineers' beliefs were one thing; nature was another. If engineers in the 1870s believed they could stop floods, blast away mountains, or tunnel miles underground without consequence, many had changed their tune only half a century later. With rare insight, Hammond admitted in 1921, "One cannot juggle with the forces of nature. All attempts to cheat nature and get round her laws result in disaster."[25] Engineers' assumptions about people and nature, in turn, influenced their ideas about their responsibilities toward society. Their thinking and personalities combined with host countries' institutional arrangements to make a common world destiny an increasingly unobtainable vision.

Engineers avowed goodwill when working in other countries yet often acted in service to personal interest. They believed in technical innovation and refuted the status quo. They realized technology's impact on economies but did not always comprehend its social ramifications. They understood the practical applications of their work but rarely made large-scale policy decisions. They entered countries looking through rose-colored lenses and left when the going got rough. They were often inquisitive, didactic, and profoundly incorrect on important social issues. In the end, they blamed the environment, the people, and the politics, thus absolving themselves of responsibility for technology's social outcomes. Nonetheless, California engineers' projects and ideologies also embodied fundamental truths—about nature, about markets, and most important, about themselves as members of a growing international community.[26] Hammond, engineering's poster child for this era, admitted, "Perhaps I have made a superman of my engineer." And, perhaps rightly, so he did.[27]

## Introduction

1 Benedict, "The Anthropology of World's Fairs," in *Anthropology*, 2–60.
2 Gordon, *What We Saw*, 12; Todd, *Story*, 4:158–62; Macomber, *Jewel City*, 146–53; Starr, *Americans*, 302–6; Panama Pacific International Exposition Company, Mining Week Committee, *Souvenir Program*; U.S. Bureau of Mines, *Exhibits*; "United States Geological Survey," 383–84; "Exhibits of the Pelton Water Wheel Co., Panama-Pacific International Exhibition, 1915," Panama Pacific International Exposition Pamphlets, folder 10.
3 C. N. Bennett, "Nation's Herculean Task," 506–7; Baxter, *Panama Canal*, 5–10, 29; Claybourn, *Dredging*, 23, 27, 36, 96; see also I. Bennett, Hammond, and Lennox, *History*.
4 Gordon, *What We Saw*, 83; Schlereth, *Victorian America*, 297; Todd, *Story*, 1:150–51.
5 "The World's Visit," *San Francisco Examiner*, August 9, 1914.
6 Marx and Engels, *Manifesto*, in Tucker, *Marx-Engels Reader*, 475, 477.
7 Hobsbawm, *Age of Capital*, 29–47.
8 Hobsbawm, *Age of Empire*, 26–33.
9 Ibid.
10 Morison, "Address," 469–70, 483; Gerber, Prout, and Schenider, "Memoir."
11 Morison, "Address," 469–70.
12 Moore, "San Francisco and the Exposition," 196; "Panama-Pacific Exposition Comes to an End," 1138.
13 Moore, "San Francisco and the Exposition," 196–98; Panama-Pacific International Exposition Company, Committee on Exploration and Publicity, *San Francisco*, 15–18.
14 Skiff, "Neighborhood," 625.
15 Gordon, *What We Saw*, 47.
16 Skiff, "Neighborhood," 623; Kinsley, "Panama-Pacific Exposition," 20.
17 Worster, *Nature's Economy*, 170–87.
18 Hobsbawm, *Age of Empire*, 31–32.
19 See J. B. Walker, "1915 Exposition."
20 Macomber, *Jewel City*, 52.
21 Ibid., 154.
22 Ibid., 154–68.
23 Todd, *Story*, 5:8.

24 Springer and Springer, "Water-Way," 547.

25 Rodgers, *Atlantic Crossings*, 7; Curti and Birr, *Prelude*, 16–24, 140–52.

26 Rodgers, *Atlantic Crossings*, 3.

27 Ibid.

28 Wolf, *Europe*, 391.

29 Newell, "Awakening," 568.

30 Ibid.; Frederick H. Newell, "The Engineer in Public Service," in Layton, *Revolt*, 118.

31 Layton, *Revolt*, 29.

32 Newell, "Awakening," 568.

33 Skiff, "Neighborhood," 625; J. H. Hammond, *Autobiography*, 2:735.

34 Headrick, *Tentacles*, 9–13.

35 J. H. Hammond, "Engineer in Public Life," 115.

36 Veblen, *Engineers*; Mumford, *Technics*, 219–20; Bledstein, *Culture*, 1–45, Layton, *Revolt*, 64.

37 See Friedman, *World*, which argues that we have now entered a stage called Globalization 3.0.

38 J. H. Hammond, *Autobiography*, 2:735.


### Chapter One

1 Jackson, *Building*, 35–37; Kahrl, *California Water Atlas*, 16.

2 Quoted in Gerald D. Nash, *State Government*, 33.

3 Ibid., 40–41.

4 McWilliams, *California*, 25, 117–18.

5 Donald J. Pisani, "The Origins of Western Water Law: Case Studies from Two California Mining Districts," in *Water, Land, and Law*, 24–37.

6 Hall, *Report of the State Engineer to the Legislature*, pt. 4, 4; Pisani, *From the Family Farm*, 175–76.

7 Newell quoted in H. M. Wilson, "Irrigation in India," *Transactions*, 256, 259.

8 Victoria, Royal Commission, *First Progress Report*; Worster, *Rivers*, 147.

9 Deakin, *Irrigated India*, 148.

10 Victoria, Royal Commission, *First Progress Report*, 13–14, 22–23.

11 *Eighth Census*, 662; *Transactions of the State Agricultural Society, 1866–1867*, 548; *Transactions of the State Agricultural Society, 1870–1871*, 172, in Pisani, *From the Family Farm*, 102–3.

12 Liebman, *California Farmland*, 16–17; Gerald D. Nash, *State Government*, 64; Pisani, *From the Family Farm*, 286–88.

13 Pisani, *From the Family Farm*, 105–6.

14 San Joaquin and King's River Canal and Irrigation Company, *Agricultural Lands*, 7; U.S. Board of Commissioners, *Engineers and Irrigation*, 9; Pisani, *From the Family Farm*, 105–7.

15 San Joaquin and King's River Canal and Irrigation Company, *Report*, 13–15; Brereton, *Extracts*, 5; San Joaquin and King's River Canal and Irrigation Company, *Great San Joaquin and Sacramento Valleys*, 1–17; Pisani, *From the Family Farm*, 105–7, 109–10; Lavender, *California*, 299–300; Lavender, *Nothing*, 353–55.

16  Brereton, *Extracts*, 5; Pisani, *From the Family Farm*, 107; San Joaquin and King's River Canal and Irrigation Company, *Great San Joaquin and Sacramento Valleys*, 1–13; Worster, *Rivers*, 147.

17  Brereton, *Reminiscences*, 7–24; *Irrigation in California*, 16–18; Pisani, *From the Family Farm*, 107–12.

18  Brereton, *Report*, 5, 12.

19  Robert M. Brereton to William Hammond Hall, June 15, 1911, box 11, folder 30, Hall Papers, Bancroft Library; Brereton, *Extracts*, 5; Pisani, *From the Family Farm*, 108–10; Igler, "When Is a River?"; Igler, "Industrial Cowboys."

20  San Joaquin and King's River Canal and Irrigation Company, *Report*, 13; Pisani, *From the Family Farm*, 110.

21  Robert M. Brereton to William Hammond Hall, June 15, 1911, box 11, folder 30, Hall Papers, Bancroft Library; "Proposed Aid from Congress," in San Joaquin and King's River Canal and Irrigation Company, *Great San Joaquin and Sacramento Valleys*, 27–29; Pisani, *From the Family Farm*, 108–13.

22  Brereton, *Reminiscences*, 27; U.S. Board of Commissioners, *Engineers and Irrigation*, 14; Pisani, *From the Family Farm*, 112–13.

23  Pisani, *From the Family Farm*, 112.

24  San Joaquin and King's River Canal and Irrigation Company, *Report*, 32–33; Pisani, *From the Family Farm*, 112–13.

25  Pisani, *From the Family Farm*, 113

26  Ibid.; Worster, *Rivers*, 147.

27  Pisani, *From the Family Farm*, 154–55.

28  U.S. Board of Commissioners, *Engineers and Irrigation*, 19–20.

29  Lewis, *George Davidson*, M. L. Smith, *Pacific Visions*, 18–27; Davenport, *Biographical Memoir*; Wagner, "George Davidson," 299–320; Pisani, *To Reclaim*, 133–38; Worster, *Rivers*, 147.

30  *Pacific Rural Press*, October 18, 1873, 248, quoted in Pisani, *From the Family Farm*, 117–18.

31  Gerald D. Nash, *State Government*, 63–80.

32  Tyrrell, *True Gardens*, 36–55.

33  *Report of the Board of Commissioners on the Irrigation of the San Joaquin, Tulare, and Sacramento Valleys of the State of California*, 25, 39, 40; Pisani, *From the Family Farm*, 114; Worster, *Rivers*, 147.

34  George Davidson to Superintendent, U.S. Coast Geodetic Survey, June 23, 1873, letterbook, vol. 26, Davidson Papers; Pisani, *From the Family Farm*, 113–14.

35  Pisani, *From the Family Farm*, 115.

36  Ibid., 116–17.

37  Gerald D. Nash, *State Government*, 208–9; Pisani, *From the Family Farm*, 117–18.

38  George Davidson, "Opinions of the U.S. Commissioners of Irrigation in 1873," 29, 36–37, box 7, Davidson Papers.

39  Pisani, *From the Family Farm*, 116; Davidson, "Opinions," 36–37.

40  George Davidson, "Irrigation: No. I: Two Papers Read before the Legislature of California, Jan. 15 and 16, 1878," 11, box 7, Davidson Papers.

41  Davidson, "Opinions," 36–37.

42  Pisani, *From the Family Farm*, 119.

43 Robert M. Brereton to William Hammond Hall, June 15, 1911, box 11, folder 30, Hall Papers, Bancroft Library; Pisani, *From the Family Farm*, 118–19.

44 U.S. Board of Commissioners, *Engineers and Irrigation*, 31; Pisani, *From the Family Farm*, 118–19.

45 Hall, "Statement," 208.

46 Davidson, "Irrigation: No. I," 3.

47 Deakin, *Irrigated India*, 148.

48 Islam, *Irrigation*, 138–48; Awasthi, *History of Development*, 47–48; K. N. Chaudhuri, "Foreign Trade and Balance of Payments (1757–1947)," in *Cambridge Economic History*, ed. Kumar, 877; Ali, *Punjab*, 237–43.

49 See accounts of Davidson's visit to India in Pisani, *To Reclaim*, 133–38; Worster, *Rivers*, 147–48.

50 C. E. Norton, "Irrigation in India," 439; "Irrigation in India," *Forestry and Irrigation*, 276–77; "Extensive Irrigating Project," 50; "Irrigation in India," *Scientific American* (1893), 89–90.

51 "Irrigation Works," 128.

52 U.S. Senate, *Report*, 249–312; Hinton, *Irrigation*, 265–328.

53 Worster, *Rivers*, 148.

54 Ibid.; Pisani, *To Reclaim*, 134–37.

55 U.S. Coast and Geodetic Survey, *Irrigation and Reclamation*, 8; Davidson, "Application"; Worster, *Rivers*, 147–51; Pisani, *To Reclaim*, 136.

56 H. M. Wilson, "Irrigation in India," *Transactions*, 240, 246–47.

57 U.S. Coast and Geodetic Survey, *Irrigation and Reclamation*, 31–32.

58 Davidson, "Application," 107; George Davidson letterbook, vol. 30, box 2, Davidson Papers; U.S. Coast and Geodetic Survey, *Irrigation and Reclamation*, 40–47, 8–12.

59 "Irrigation in India," *Scientific American* (1893): 89–90; H. M. Wilson, "Irrigation in India," *Transactions*, 225–26; Brown, "Irrigation," 3–31; "Irrigation in India," *Forestry and Irrigation*, 276.

60 U.S. Coast and Geodetic Survey, *Irrigation and Reclamation*, 8–9; Pisani, *To Reclaim*, 136–37.

61 U.S. Coast and Geodetic Survey, *Irrigation and Reclamation*, 39, 11, 57.

62 Wescoat, "Water Rights," 19–20, 28–29; Pisani, *To Reclaim*, 136.

63 Davidson, "Application," 105–9; "Irrigation Works," 128; "Extensive Irrigating Project," 50; "Irrigation in India," *Scientific American* (1883), 279; "Irrigation in India," *Scientific American* (1893): 89–90; U.S. Coast and Geodetic Survey, *Irrigation and Reclamation*, 38–39; Pisani, *To Reclaim*, 136–37; Worster, *Rivers*, 148.

64 See Wittfogel, *Oriental Despotism*; Worster, *Rivers*, 22–30.

65 David Gilmartin, "Models of the Hydraulic Environment: Colonial Irrigation, State Power, and Community in the Indus Basin," in *Nature, Culture, Imperialism*, ed. Arnold and Guha, 210–36; Worster, *Rivers*, 21–50; Wolpert, *New History*, 238; Harris, *Irrigation*, 243–44; Headrick, *Tools*, 180–91.

66 Whitcombe, *Agrarian Conditions*, 24.

67 Shariff, *Development*, 21–23; Jha, *Irrigation*, 57; Wescoat, "Water Rights," 19.

68 Davidson, "Application," 107.

69 U.S. Coast and Geodetic Survey, *Irrigation and Reclamation*, 11; Anstey, *Economic*

Development, 129, 526; Porter, *Lion's Share*, 41; Wolpert, *New History*, 238; Harris, *Irrigation*, 9; Worster, *Rivers*, 148–51.

70  Buckley, *Irrigation Works*, 92–101.

71  H. M. Wilson, "Irrigation in India," *Transactions*, 240, 246–47.

72  Brown, "Irrigation."

73  H. M. Wilson, "Irrigation in India," *Transactions*, 221; Buckley, *Irrigation Works*, 18; Bellasis, *Punjab Rivers*, 1–3; Brown, "Irrigation," 11–13.

74  Ali, *Punjab*, 13.

75  Deakin, *Irrigated India*, 123.

76  Brown, "Irrigation," 12.

77  Harris, *Irrigation*, 49; H. M. Wilson, "Irrigation in India," *Transactions*, 237; Ali, *Punjab*, 175.

78  Islam, *Irrigation*, 17; Harris, *Irrigation*, 10, 49–52; Buckley, *Irrigation Works*, 2–6, 17–18; Worster, *Rivers*, 148–51.

79  Gilmartin, "Scientific Empire"; H. M. Wilson, "Irrigation in India," *Transactions*, 218–19; Whitcombe, *Agrarian Conditions*, 10.

80  Wescoat, "Water Rights," 20–22; Gilmartin, "Scientific Empire"; Ali, *Punjab*, 10–12.

81  Worster, *Rivers*, 151–52; Gilmartin, "Scientific Empire," 1141; Ali, *Punjab*, 13–14, 99–104, 161, 239; Whitcombe, *Agrarian Conditions*, 14–17, 118–19; Islam, *Irrigation*, 80–82.

82  U.S. Coast and Geodetic Survey, *Irrigation and Reclamation*, 38; Whitcombe, *Agrarian Conditions*, 85–87; Worster, *Rivers*, 148–52.

83  Davidson, "Application," 106.

84  Whitcombe, *Agrarian Conditions*, 285; Worster, *Rivers*, 153.

85  U.S. Coast and Geodetic Survey, *Irrigation and Reclamation*, 39, 11, 57; Worster, *Rivers*, 152.

86  U.S. Coast and Geodetic Survey, *Irrigation and Reclamation*, 39, 11, 57; J. T. Walker to George Davidson, March 5, 1875, March 21, 1879, box 14, Davidson Papers.

87  U.S. Coast and Geodetic Survey, *Irrigation and Reclamation*, 40–47; Pisani, *To Reclaim*, 133–36.

88  U.S. Coast and Geodetic Survey, *Irrigation and Reclamation*, 39.

89  Ibid., 40, 69; Pisani, *To Reclaim*, 136.

90  U.S. Coast and Geodetic Survey, *Irrigation and Reclamation*, 40, 69.

91  Ibid.

92  Pisani, *From the Family Farm*, 162–63

93  Ibid., 163; Kelley, *Gold versus Grain*.

94  Kelley, *Gold versus Grain*, 14–15.

95  Davidson, "Irrigation: No. I," 2–3, 5.

96  George Davidson, lecture on irrigation before the California legislature, January 15–16, 1878, 34, Davidson Papers.

97  Ibid.

98  Campbell Polson Berry to George Davidson, January 19, 1878, March 28, 1878, box 2, Davidson Papers; Pisani, *From the Family Farm*, 166–67.

99  Davidson, lecture; Pisani, *From the Family Farm*, 167, 154–90; Kelley, *Gold versus Grain*.

100  William Hammond Hall, "Drainage and Debris Work of 1878–81: First Letter: The

Origin and Conditions of the Investigations," MSS 913/I/10, Hall Papers, California Historical Society, 5.

101 Pisani, *From the Family Farm*, 166, 138–39, 158–59; U.S. Board of Commissioners, *Engineers and Irrigation*, 35.

102 William Hammond Hall, "Recollections of Early California Engineering," box 7, folder 14, Hall Papers, Bancroft Library, 41–46; Pisani, *From the Family Farm*, 143–46, 166–67.

### Chapter Two

1 William Hammond Hall, "Recollections of Early California Engineering," 2, n.d., box 7, folder 14, William Hammond Hall Papers, Bancroft Library; Starr, *Material Dreams*, 7; Gerald D. Nash, *State Government*, Pisani, *From the Family Farm*, 154–90; Teele, *State Engineer*, 8.

2 Hall, "Recollections," 7–8; Pisani, *From the Family Farm*, 167.

3 Hall, "Recollections," 42–45; Pisani, *From the Family Farm*, 140–43.

4 Hall, "Recollections," 47–51.

5 William Hammond Hall, "Drainage and Debris Work of 1878–1881: First Letter: The Origin and Conditions of the Investigations," 5, MSS 913/I/10, Hall Papers, California Historical Society; Hall, "Statement," 208; Pisani, *From the Family Farm*, 166–67.

6 Hall, "Drainage and Debris Work: First Letter," 3.

7 Ibid., 2–3.

8 William Hammond Hall to Hamilton Smith, October 13, 1882, folder 9, box 2, Hall Papers, Bancroft Library.

9 Hall, "Drainage and Debris Work: First Letter," 4; Pisani, *From the Family Farm*, 164–66.

10 Hall, "Statement," 209; Pisani, *From the Family Farm*, 168.

11 Hall, *Report of the State Engineer to the Legislature*, pt. 3, 14–15.

12 Ibid., 15–20, 27–31, 49–50, 18–19, appendix A, 49, appendix B, 68.

13 Kelley, *Gold versus Grain*, 147–50; Pisani, *From the Family Farm*, 168–71.

14 Correspondence between Hall and Eads, box 12 folder 17, Hall Papers, Bancroft Library; correspondence between Hall and Mendell, box 15, folders 56–57, Hall Papers, Bancroft Library; Hall, "Recollections," 20; Sackett, *Engineer*, 166–67.

15 Hall, *Proceedings*, 12.

16 Hall, *Memorandum*, 3, 8–9, 11.

17 Pisani, *From the Family Farm*, 168, 172; Kelley, *Gold versus Grain*, 99–100; "The Debris Question from an Engineer's Standpoint," *Daily Alta California*, April 6, 1879; Hall, *Report of the State Engineer to the Legislature*, pt. 3, 31, 42–43.

18 Hall, *Proceedings*, 18–19.

19 Kelley, *Gold versus Grain*, 139–44, 150–52; Pisani, *From the Family Farm*, 168–71.

20 William Hammond Hall, "The Drainage and Debris Works of 1878–81: Ninth Letter: Brief Summary of the Views of the State and Consulting Engineers. The Outcome of the Work," MSS 913/I/10, Hall Papers, California Historical Society, 3–4; Pisani, *From the Family Farm*, 172.

21  Hall, *Second Report*, pt. 2, 3:15–27; Kelley, *Gold versus Grain*, 154; Pisani, *From the Family Farm*, 172–73; Hall, "Drainage and Debris Works: Ninth Letter," 7.

22  Hall, "Drainage and Debris Works: Ninth Letter," 7; Hall, *Second Report*, pt. 4, 24.

23  Hall, *Proceedings*, 4–7; Hall, "Drainage and Debris Works: Ninth Letter," 1–2; Pisani, *From the Family Farm*, 169, 173.

24  Pisani, *From the Family Farm*, 173–74.

25  Correspondence between Hall and Davidson, boxes 4, 13, Davidson Papers.

26  Hall, *Report of the State Engineer to the Legislature*, pt. 3; Hall, *Irrigation Question: Memorandum*, 9–11; Kelley, *Battling*, 203; Hall, *Second Report*, pt. 4, 7.

27  Hall, *Fifth Progress Report*, 8; Hall, "Irrigation Principles I," 221; Hall, "Irrigation Principles II," 253.

28  Hall, *Irrigation Question: Memorandum*, 10.

29  Pisani, *From the Family Farm*, 180–81.

30  Hall, *Second Report*, pt. 4, 11–16, 20–31; Pisani, *From the Family Farm*, 176–77.

31  Pisani, *From the Family Farm*, 179–80.

32  Hall, "Drainage and Debris Works: Ninth Letter," 8; Pisani, *From the Family Farm*, 174.

33  *Water and Land* 1, no. 2 (1886); Hall, *Fifth Progress Report*, 14; Pisani, *From the Family Farm*, 182–84.

34  Hall, "Drainage and Debris Work: First Letter," 9–10; Starr, *Material Dreams*, 10–11.

35  Kelley, *Battling*, 217–19; Kelley, *Gold versus Grain*.

36  Hundley, *Great Thirst*, 91–97; Pisani, *From the Family Farm*, 191–249; Igler, "When Is a River?"

37  Hundley, *Great Thirst*, 97–102; Pisani, *From the Family Farm*, 250–82.

38  Hall, "Drainage and Debris Works: Ninth Letter," 8; Pisani, *From the Family Farm*, 184–85.

39  Pisani, *From the Family Farm*, 188–90; Teisch, "William Hammond Hall," 81–98.

40  *Choice Bit*.

41  Smythe, "Real Utopias," 605–6; Pisani, *From the Family Farm*, 80–82.

42  Pisani, *From the Family Farm*, 82–83; Hundley, *Great Thirst*, 102–3.

43  Smythe, "Real Utopias," 607; Hall, *Irrigation in [Southern] California*, 209–12, 234–40; Hundley, *Great Thirst*, 103–4.

44  Pisani, *From the Family Farm*, 80–82; Smythe, "Real Utopias," 606; Smythe, *Conquest*, 94–97.

45  Smythe, *Conquest*, 131.

46  Hall, *Irrigation in [Southern] California*, 113–14.

47  "Fair Etiwanda," *Fruit Grower*, December 11, 1882.

48  J. A. Alexander, *Life*, 34–36, 51–52; Pisani, *From the Family Farm*, 281–82; Hall, *Irrigation in [Southern] California*, 335–38.

49  J. A. Alexander, *Life*, 34–36, 51–52; Pisani, *From the Family Farm*, 281–82.

50  Hall, *Irrigation in [Southern] California*, 335; *Souvenir from Etiwanda*.

51  J. A. Alexander, *Life*, 37–38; Hill, *Water into Gold*, 44.

52  Gentilcore, "Ontario, California," 81.

53  Widney, *Ontario*, 5; Chaffey Brothers, *Australian Irrigation Colonies*, 107; J. A. Alexander, *Life*, 48; Hill, *Water into Gold*, 45.

54 *Choice Bit.*
55 Conley, *Dreamers and Dwellers*, 49–50.
56 Gentilcore, "Ontario, California," 85; Widney, *Ontario*, 9–10.
57 Quoted in California Excursion Association, *Southern California*, 62.
58 *Life and a Living.*
59 Fortier, "Field Work," 375.
60 *Etiwanda*, 11.
61 Moeller, *Ontario-Cucamonga-Etiwanda Colonies.*
62 Chaffey Brothers, *Australian Irrigation Colonies*, 107.
63 "Thriving Settlements," *Ontario Fruit Grower*, August 29, 1883.
64 *Life and a Living.*
65 J. A. Alexander, *Life*, 51; Trask, "Irrigation System."
66 Gentilcore, "Ontario, California," 86; D. H. Walker, *Ontario, California*, 3.
67 Gentilcore, "Ontario, California," 84; Trask, "Irrigation System," 177; J. A. Alexander, *Life*, 55, 67.
68 Osborne, *Ontario Colony.*
69 Conley, *Dreamers and Dwellers*, 72; Hickcox, *History*, 7, 54.
70 Vincent, *Colonisation*, 20; Osborne, *Ontario Colony*; Moeller, *Ontario-Cucamonga-Etiwanda Colonies.*
71 Conley, *Dreamers and Dwellers*, 4; Moeller, *Ontario-Cucamonga-Etiwanda Colonies*; Osborne, *Ontario Colony*; D. H. Walker, *Ontario, California*, 23.
72 D. H. Walker, *Ontario, California*, 5.
73 Chaffey Brothers, *Australian Irrigation Colonies*, 111.
74 J. A. Alexander, *Life*, 59–69.
75 Pisani, *From the Family Farm*, 283–85; Pomeroy, *Pacific Slope*, 102, 110–11.
76 Pisani, *From the Family Farm*, 286–89; Pisani, "Water Law Reform," 296.
77 Pisani, *From the Family Farm*, 289–90.
78 Pisani, *To Reclaim*, 273.
79 Hundley, *Great Thirst*, 111–12, 117–18.
80 S. P. Hays, *Conservation*, 265.
81 Hundley, *Great Thirst*, 115–16; Pisani, *From the Family Farm*, 301.
82 Lippincott, "Reclamation Service"; Pisani, *From the Family Farm*, 301–8; Newell, "Work," 346–47.
83 Worster, *Rivers*, 194–95.
84 Ibid., 195–96; Hundley, *Great Thirst*, 204–9; Starr, *Material Dreams*, 21.
85 Wright, "How the Commercial Organizations"; Grunsky, "Lower Colorado River," 3–4; Hundley, *Great Thirst*, 204–9.
86 deBuys and Myers, *Salt Dreams*, 78–79; Grunsky, "Lower Colorado River," 20–21; Worster, *Rivers*, 196; Hundley, *Great Thirst*, 205.
87 J. A. Alexander, *Life*, 293; deBuys and Myers, *Salt Dreams*, 78–81; Grunsky, "Lower Colorado River," 20–22; Pisani, *From the Family Farm*, 309–11.
88 deBuys and Myers, *Salt Dreams*, 78–81, 88; Hundley, *Great Thirst*, 204–9; J. A. Alexander, *Life*, 270–312; Grunsky, "Lower Colorado River," 4; Worster, *Rivers*, 196.
89 Lippincott, "Reclamation Service," 169; Lippincott, "General Outlook," 351; Pisani, *From the Family Farm*, 309–11.
90 Starr, *Material Dreams*, 34; deBuys and Myers, *Salt Dreams*, 91; Hundley, "Poli-

tics of Reclamation," 302; Lippincott, "Reclamation Service," 169; Pisani, *From the Family Farm*, 313–15.

91  deBuys and Myers, *Salt Dreams*, 91–95; Hundley, *Great Thirst*, 204–9.

92  deBuys and Myers, *Salt Dreams*, 91–95, 101–6; Grunsky, "Lower Colorado River," 2–3; Worster, *Rivers*, 196–97; Hundley, *Great Thirst*, 205–6.

93  deBuys and Myers, *Salt Dreams*, 110–19; Lavender, *California*, 319–24; J. A. Alexander, *Life*, 270–312; Worster, *Rivers*, 197.

94  deBuys and Myers, *Salt Dreams*, 120–21.

95  Worster, *Rivers*, 201; Pisani, *From the Family Farm*, 318, 319; Hundley, *Great Thirst*, 206.

96  Pisani, *To Reclaim*, 324–25, 273; F. Adams, "Frank Adams," 105–6; Hundley, *Great Thirst*, 237.

97  Hundley, *Great Thirst*, 248–68; 209–23; Worster, *Rivers*, 207–9; Pisani, *From the Family Farm*, 309–12; deBuys and Myers, *Salt Dreams*, 126.


## Chapter Three

1  Victoria, Royal Commission, *First Progress Report*, 34; Dunlap, *Nature*, 3.

2  Tyrrell, *True Gardens*, 4–9.

3  Joseph M. Powell, *Historical Geography*, 1–43.

4  Deakin, *Water Supply*, 9; Deakin, "Irrigation in Australia," 86.

5  Joseph M. Powell, *Watering*, 37–53, 81–90; Joseph M. Powell, *Historical Geography*, 19–55.

6  Deakin, *Water Supply*, 21; Victoria, Rivers and Water Supply Commission, *Irrigation and Water Supply Development*, 8; Deakin, "Irrigation in Australia," 86.

7  Spence, *Mining Engineers*, 296–300.

8  Joseph M. Powell, *Historical Geography*, 41–43.

9  Ibid., 19–55; Joseph M. Powell, *Watering*, 60–84.

10  Ballantyne, *Our Colony*, 25–26; East, *Victorian Water Law*, 3–4; Perkins, *Melbourne Illustrated*, 25–26.

11  A. S. Murray, *Twelve Hundred Miles*, 26; U.S. Senate, *Report*, 298–307.

12  Quoted in McCoy, *Victorian Irrigation*, 11.

13  Aplin, Foster, and McKernan, *Australians*, 109–10; Tyrrell, *True Gardens*, 154.

14  See Tyrrell, *True Gardens*, 4–12.

15  Deakin, "Irrigation in Australia," 86; Tyrrell, *True Gardens*, 122; McCoy, *Victorian Irrigation*, 4–5; Worster, *Rivers*, 152–52.

16  William Hammond Hall to Hugh McColl, October 24, 1884, box 3, folder 1, Hall Papers, Bancroft Library; Tyrrell, *True Gardens*, 122–23.

17  John L. Dow to George Stoneman, July 16, 1884, ms. 915, box 1, folder 1, Hall Papers, California Historical Society; Tyrrell, *True Gardens*, 123.

18  Deakin quoted in Hill, *Water into Gold*, 42.

19  Alfred Deakin, travel letters, March 3, 14, 24, 1885, Deakin Papers.

20  Victoria, Royal Commission, *First Progress Report*, 24–29, 101–2, 11–12.

21  Joseph M. Powell, "Enterprise and Dependency: Water Management in Australia," in *Ecology and Empire*, ed. Griffiths and Robin, 110–11; Tyrrell, *True Gardens*, 123–25.

22  Victoria, Royal Commission, *Further Progress Report*, 45; Victoria, Royal Commission, *First Progress Report*, 30; Tyrrell, *True Gardens*, 126–27.

23  Hill, *Water into Gold*, 47; Victoria, Mildura Settlement, *Report*, vi; "Synopsis of Evidence," in Victoria, Mildura Settlement, *Report*, xci–xcii; J. A. Alexander, *Life*, 52; Victoria, Royal Commission, *Further Progress Report*, 34; Tyrrell, *True Gardens*, 143.

24  Victoria, Royal Commission, *First Progress Report*, appendix B, 21. A summary of Deakin's report is in U.S. Senate, *Report*, 296–320.

25  Alfred Deakin to unidentified, April 6, 1885, Deakin Papers; William Hammond Hall to Alfred Deakin, March 28, 1885, box 3, folder 1, Hall Papers, Bancroft Library.

26  Victoria, Royal Commission, *First Progress Report*, 111–12; Hall, *Irrigation Question: Memorandum no. 1–2*, 9; William Hammond Hall to Alfred Deakin, April 10, 11, 12, 1885, box 3, folder 1, Hall Papers, Bancroft Library; Tyrrell, *True Gardens*, 150–51.

27  Victoria, Royal Commission, *First Progress Report*, 46, 22–23; Worster, *Rivers*, 146, 150–51.

28  Tyrrell, *True Gardens*, 124–25.

29  Lavender, *California*, 295–97; Starr, *Americans*, 134–41; Tyrrell, *True Gardens*, 131–33; Dow, *Our Land Acts*, 10–12.

30  Tyrrell, *True Gardens*, 132–33.

31  Victoria, Royal Commission, *First Progress Report*, 106–9; Tyrrell, *True Gardens*, 126–27, 141–43; Kluger, *Turning on Water*, 59–60.

32  Hall, *Irrigation Question: Memorandum no. 1–2*, 8; Deakin, *Guide*, 5–107; East, *Victorian Water Law*, 5–6; Victoria, Rivers and Water Supply Commission, *Irrigation Trust Blue Books*.

33  McCoy, *Victorian Irrigation*, 13; *New Eldorado*, 5.

34  McCoy, *Victorian Irrigation*, 13–18; Joseph M. Powell, *Watering*, 118; Tyrrell, *True Gardens*, 141–43.

35  Victoria, Rivers and Water Supply Commission, *Irrigation and Water Supply*, 9–10; Rutherford, "Interplay," 114–20; Tyrrell, *True Gardens*, 141–43.

36  Joseph M. Powell, *Watering*, 120–22.

37  Donald J. Pisani, "Reclamation and Social Engineering," in *Water, Land, and Law*, 180–94; Layton, *Revolt*, 58–60.

38  "The Proposed Irrigation Settlement," *Register*, February 15, 1887; Stuart Murray, *Irrigation*, 5; Tyrrell, *True Gardens*, 144.

39  L. Peacock, "George Chaffey: Famous Irrigationist," *Sydney Morning Herald*, April 18, 1934; Perkins, *Melbourne Illustrated*, 24, 33; Chaffey Brothers, *Australian Irrigation Colonies*, 98.

40  Chaffey Brothers, *Australian Irrigation Colonies*, 5; Victoria, *Murray River Waters*.

41  Victoria, Mildura Settlement, *Report*, vii; "The Chaffey Brothers' Irrigation Colony," *Daily Telegraph*, July 11, 1884; Myers, *Irrigation*, 10–11; Tyrrell, *True Gardens*, 146.

42  Perkins, *Melbourne Illustrated*, 32; Joseph M. Powell, *Watering*, 121; Tyrrell, *True Gardens*, 147.

43  Voullaire, *Mildura Irrigation Settlement*, 19.

44  Vincent, *Colonisation*, 2.

45  Woehlke, "Food First," 5.
46  Vincent, *Colonisation*, 2.
47  Chaffey Brothers, *Australian Irrigation Colonies*, 3.
48  Myers, *Irrigation*, 3–4.
49  Burton, *How to Get Rich*, 12–14.
50  Myers, *Irrigation*, 10–11; Hill, *Water into Gold*, 78–79; "Irrigation Farms in Australia," in *Renmark Newspaper Slips*, ed. Thomson and Wells, 169.
51  Chaffey Brothers, *Australian Irrigation Colonies*, 83; Tyrrell, *True Gardens*, 145.
52  Burton, *How to Get Rich*, 17; Myers, *Irrigation*, 52; Tyrrell, *True Gardens*, 145.
53  "Visit to Mildura by the Right Rev. Dr. Thornton, Bishop of Ballarat," *Ballarat Star*, April 3, 1888.
54  Myers, *Irrigation*, 35.
55  Stephen Cureton, "The Value and Importance of Irrigation in Australia," in Vincent, *Colonisation*, 48; Stuart Murray, *Irrigation*, 6; "Visit to Mildura"; Tyrrell, *True Gardens*, 145; Stuart Murray, *Irrigation*, 6.
56  Stuart Murray, *Irrigation*, 6; "Visit to Mildura."
57  "Irrigation of Australia," 189.
58  A. S. Murray, *Twelve Hundred Miles*, 23–24; Myers, *Irrigation*, 18–19.
59  Sir Henry Parkes, Report of Speech, September 1887, in Vincent, *Colonisation*, 1.
60  Tyrrell, *True Gardens*, 145–46; "Visit to Mildura"; Chaffey Brothers, *Australian Irrigation Colonies*, 73.
61  Wells, *Paddle Steamers*, 135, 189, 193; J. A. Alexander, *Life*, 184–85.
62  Vincent, *Colonisation*, 34; Tyrrell, *True Gardens*, 146–48.
63  J. A. Alexander, *Life*, 188–90; Victoria, Mildura Settlement, *Report*, xii–xvi; Tyrrell, *True Gardens*, 146–47.
64  Victoria, Mildura Settlement, *Report*, viii; Tyrrell, *True Gardens*, 147.
65  Stuart Murray, "Mildura Settlement: Report by the Chief Engineer of Water Supply," in Mildura Royal Commission, *Mildura Irrigation Settlement*, 6–7.
66  "Synopsis of Evidence," lxxii, xlvii.
67  Kershner, "George Chaffey," 119.
68  "Synopsis of Evidence," xlvii.
69  "Minutes of Evidence, Wednesday, 1st July, 1896," in Victoria, *Mildura Royal Commission*, 49, 169.
70  Victoria, Royal Commission, *First Progress Report*, 12–14; J. A. Alexander, *Life*, 89–90, 104–5, 114–15; Kershner, "George Chaffey," 120.
71  Ballantyne, *Our Colony*, 58; Vamplew, *Australians*, 166, 168; Branagan, *Railway Policy*, 1–3; "Minutes of Evidence, Friday, 29th May, 1896," in Victoria, Mildura Settlement, *Report*, 38; Tyrrell, *True Gardens*, 148.
72  "An Interview with the Messrs. Chaffey," *Melbourne Argus*, cited in Vincent, *Colonisation*, 59; Tyrrell, *True Gardens*, 148.
73  Victoria, Mildura Settlement, *Report*, xx.
74  Vamplew, *Australians*, 26, 34.
75  Victoria, Mildura Settlement, *Report*, xvi.
76  Chaffey Brothers, *Australian Irrigation Colonies*, 35; Vamplew, *Australians*, 88, 98; "Mines of West Australia," 39; "Australian Gold," 149; "Australian Gold Production," 300–301.

77 Kershner, "George Chaffey," 118; J. A. Alexander, *Life*, 168–86; Alfred Deakin to William B. Chaffey, June 30, 1898, series 10, item 80, Deakin Papers.

78 Stuart Murray to Alfred Deakin, October 1, 1907, series 10, item 93–95, Deakin Papers.

79 Wells, *Paddle Steamers*, 202; Tyrrell, *True Gardens*, 148.

80 Joseph M. Powell, "Elwood Mead," 330–31.

81 Rutherford, "Interplay," 125.

82 Elwood Mead, "Systematic Aid to Settlers the First Need in Irrigation Development," 3, address delivered at the Irrigation Conference, Denver, April 9, 1914, Mead Papers, Water Resources Center Archives.

83 Rutherford, "Interplay," 124–25.

84 Andrews, "Irrigation," 14–15; Worster, *Rivers*, 130–31; Tyrrell, *True Gardens*, 152.

85 McCoy, *Victorian Irrigation*, 19–20; Victoria, Rivers and Water Supply Commission, *Irrigation and Water Supply*, 10–13.

86 Baker, "Elwood Mead," 4; Kluger, *Turning on Water*, 3–26.

87 Conkin, "Vision," 88; S. P. Hays, *Conservation*, 243–44; F. Adams, "Frank Adams," 149–53; Kluger, *Turning on Water*, 27–40.

88 S. P. Hays, *Conservation*, 1–4, 261–76; Joseph M. Powell, "Elwood Mead," 328–29; Kluger, *Turning on Water*, ix–xiii, 14–40; F. Adams, "Frank Adams," 98–103; Pisani, *From the Family Farm*, 342; Donald J. Pisani, "State vs. Nation: Federal Reclamation and Water Rights in the Progressive Era," in *Water, Land, and Law*, 38–49.

89 Kluger, *Turning on Water*, 39–40, 64–65.

90 Rutherford, "Interplay," 123; Tyrrell, *True Gardens*, 155–56; Kluger, *Turning on Water*, 60–61.

91 Mead, "Systematic Aid," 1; Victoria, Rivers and Water Supply Commission, *Advice*, 1; Baker, "Elwood Mead," 4–6; McCoy, *Victorian Irrigation*, 20–25; Kluger, *Turning on Water*, 64–65; Mead, "Rural Credits," 1–9.

92 "Social Effects of Close Settlement," *Irrigation in America*, 9; James H. McColl to Alfred Deakin, June 30, 1905, series 16, item 429–30, Deakin Papers; McColl, *Agriculture*; Reay, *Irrigation*; *Irrigation in America*, 4, 11.

93 Mead, "Systematic Aid," 3; "Irrigated Victoria: Advantages over America," in Gullett, *Irrigation*, 35–37; Conkin, "Vision," 88–89; John Murray, *Victorian Settlers' Guide*, 7–8; Kluger, *Turning on Water*, 65.

94 Conkin, "Vision," 88–89; Joseph M. Powell, *Historical Geography*, 53–54; Rutherford, "Interplay," 128; Kluger, *Turning on Water*, 65–67.

95 Baker, "Elwood Mead," 7–8; McCoy, *Victorian Irrigation*, 23; Rutherford, "Interplay," 127–28; Kluger, *Turning on Water*, 66.

96 Rutherford, "Interplay," 130–31.

97 *Golden Wheat Lands*, map.

98 Kluger, *Turning on Water*, 67.

99 Mead, "Systematic Aid," 3.

100 Tyrrell, *True Gardens*, 164–65.

101 Mead, *Helping Men*, 9.

102 Kluger, *Turning on Water*, 66.

103 Ibid., 72.

104 Mead, "What Australia Can Teach," 370.

105 Elwood Mead to John D. Works, December 24, 1912, box 4, Works Papers; Tyrrell, *True Gardens*, 161; Kluger, *Turning on Water*, 72–73, 85–86; Conkin, "Vision," 89.
106 See S. P. Hays, *Conservation*.
107 Mead, "What Australia Can Teach," 370; Tyrrell, *True Gardens*, 161–62.
108 Pisani, *From the Family Farm*, 441–42.
109 California, Commission on Land Colonization and Rural Credits, *Report*, 7–8, quoted in Joseph M. Powell, "Elwood Mead," 337.
110 Mead, "Tenant Farmer," 2–3; Joseph M. Powell, "Elwood Mead," 337–38.
111 Elwood Mead, "Lecture IV: Need for a Planned Rural Development," 2, November 23, 1917, box 5, Mead Papers, Bancroft Library; Mead, "New Forty-niners," 651–58, 702–3.
112 *Report of the Commission on Country Life*, 24, cited in Ellsworth, "Theodore Roosevelt's Country Life Commission," 331–32.
113 Pisani, *From the Family Farm*, 442; Tyrrell, *True Gardens*, 118–19; Starr, *Inventing*, 171–73; see also Daniels, *Politics of Prejudice*; Kluger, *Turning on Water*, 98.
114 F. Adams, "Frank Adams," 283–85; F. Adams, "Land Settlement," 199–212, 225–28; Pisani, *From the Family Farm*, 443.
115 Mead, "Lecture IV," 2; Elwood Mead, "Protection of Settlers Requires Public Examination and Approval of All Irrigated Colony Enterprises," November 29, 1916, box 16, Mead Papers, Bancroft Library; Mead, *Comparison*, 33–34; Kluger, *Turning on Water*, 86–87; Joseph M. Powell, "Elwood Mead," 337; Pisani, *From the Family Farm*, 444.
116 Behr, "Colonization Methods," 3–4; Mead, "New Forty-niners," 652; Pisani, *From the Family Farm*, 121–24.
117 Mead, *Comparison*, table 2, "Conditions of Payment, and Present Indebtedness."
118 Ibid., table 4, 61, "Rate of Interest and Time Given to Pay for Land Under Colonization Systems of Different Countries"; F. Adams, "Data."
119 Tyrrell, *True Gardens*, 164.
120 Mead, "New Forty-niners," 655–56; Joseph M. Powell, "Elwood Mead," 340.
121 Hundley, *Great Thirst*, 234–37; Kluger, *Turning on Water*, 87; California, Department of Public Works, Division of Land Settlement, *Farming*, 10–11; Pisani, *From the Family Farm*, 444.
122 Kluger, *Turning on Water*, 88–89, 92; Pisani, *From the Family Farm*, 444–45; California, Department of Agriculture, Division of Land Settlement, *Final Report*, 18.
123 Mead, "New Forty-niners," 655; Pisani, *From the Family Farm*, 444.
124 Joseph M. Powell, "Elwood Mead," 342–44; Kluger, *Turning on Water*, 87–91; Mead, *Helping Men*, 140–60; F. Adams, "Frank Adams," 287–88; Pisani, *From the Family Farm*, 444–46.
125 Mead, "New Forty-niners," 651; Pisani, *From the Family Farm*, 444–45.
126 "Developing Irrigated Land," 1014; Pisani, *From the Family Farm*, 445; Starr, *Inventing*, 170–71.
127 California, Department of Public Works, *Introduction*, 13.
128 Jones, "Ready-Made Farms," 12.
129 Starr, *Inventing*, 170–71.
130 Gertrude Matthews Shelby, "Nobody Wants to Leave This Town," *Collier's*, May 30, 1925, 26, quoted in Pisani, *From the Family Farm*, 446.

131 Mead, *Helping Men*, 140–60; Mead, *Government Aid*, 10–11; Kluger, *Turning on Water*, 88–91; Starr, *Inventing*, 171.

132 M. Adams, "Native Values," 9.

133 California, Department of Agriculture, Division of Land Settlement, *Final Report*, 4; Kluger, *Turning on Water*, 92.

134 Reid, "Agrarian Opposition," 167–79.

135 California, Department of Public Works, Division of Land Settlement, *Farming*, 4; Mead, "New Forty-niners," 702; Kluger, *Turning on Water*, 92–93; Pisani, *From the Family Farm*, 446.

136 C. P. Norton, "Where Will Our Soldiers Go?" 18; Alvin Johnson, "Land for the Returned Soldier," 218–20; "California Takes Lead," 21–22; Kluger, *Turning on Water*, 91–92; U.S. Department of the Interior, *Summary*, 5.

137 California, Department of Public Works, Division of Land Settlement, *Farming*, 5; California, Land Settlement Board, *Farm Allotments*, 3; Pisani, *From the Family Farm*, 446; Kluger, *Turning on Water*, 92–93.

138 Woehlke, "Food First," 5.

139 Mead, "Lecture IV," 1; Mead, "Tenant Farmer," 1.

140 "Australia Adopts Amortization Act with Big Success," *Denver Great Divide*, November 22, 1915; Pisani, *From the Family Farm*, 446; Kluger, *Turning on Water*, 94.

141 West, "Riveted Down," 26; Conkin, "Vision," 91; "Australia Adopts Amortization Act with Big Success," *Denver Great Divide*, November 22, 1915.

142 Kluger, *Turning on Water*, 97–101.

143 A. Shaw, "California's Farm Colonies," 400; Kluger, *Turning on Water*, 92–94; F. Adams, "Frank Adams," 288; Pisani, *From the Family Farm*, 446–47; Worster, *Rivers*, 184–85.

144 F. Adams, "Frank Adams," 289–93; Kluger, *Turning on Water*, 95–100; Pisani, *From the Family Farm*, 447–48.

145 California, Department of Agriculture, Division of Land Settlement, *Final Report*, 3; Joseph M. Powell, "Elwood Mead," 347; Conkin, "Vision," 91; Pisani, *From the Family Farm*, 447–48; Kluger, *Turning on Water*, 100.

146 California, Department of Agriculture, Division of Land Settlement, *Final Report*, 19–20.

147 See Tyrrell, *True Gardens*, 167–73.

### Chapter Four

1 Spence, *Mining Engineers*, 278–317.

2 Hall, "Rise of the Mining Industry," 19, box 6, folder 17, Hall Papers, Bancroft Library.

3 Kelley, *Gold versus Grain*, 238; "Hamilton Smith," July 28, 1900, 94.

4 Spence, *British Investments*, 8; Spence, *Mining Engineers*, 137–38; J. H. Hammond, *Autobiography*, 2:500–25; "Passing of the Pioneer," 506–7.

5 Memorandum of Association, October 26, 1886; Memorandum of Association, October 21, 1889; Register of Managers, November 27, 1886; "Abstracts," 401;

"Passing of the Pioneer," 507; Turrell and Van Helten, "Rothschilds," 182–85; de Waal, "American Technology," 81–85.

6 "Sale of Anaconda Stock," June 13, 1896, 561, September 28, 1895, 294, October 26, 1895, 389, June 13, 1896, 561; Wilkins, *Emergence*, 80.

7 *Engineering and Mining Journal*, February 18, 1905, 98; Spence, *Mining Engineers*, 282, 285.

8 Twain, *More Tramps Abroad*, 468; Spence, *Mining Engineers*, 283; Raymond, "Biographical Notice"; Malone, *Dictionary*, 9:609; "Passing of the Pioneer," 506–7.

9 "Hamilton Smith," July 14, 1900, 34; "Hamilton Smith," July 28, 1900, 94; "Passing of the Pioneer," 506–7; Malone, *Dictionary*, 9:273–75.

10 Rossiter W. Raymond to editor, *Engineering and Mining Journal*, January 25, 1896, 83.

11 Finch, "Workings," 65.

12 "Gold in South Africa," 39.

13 William Russell Quinan, "William Russell Quinan, Class of 1870: Written by Himself," 20–22, MSS Afr. s. 123, Rhodes House Library, Oxford University, England; "Witwatersrand Gold-Field," 36.

14 Mintner, *King Solomon's Mines*, 4–6.

15 Cecil Rhodes quoted in Paul Johnson, *Consolidated Gold Fields*, 30; Shillington, *History*, 53–55, 83–90, 117–22; Galbraith, *Crown and Charter*, 310–39; Hobsbawm, *Age of Empire*, 62–63.

16 Lockhart and Woodhouse, *Cecil Rhodes*, 95–113; Wheatcroft, *Randlords*; Turrell, "Rhodes."

17 Turrell and Van Helten, "Rothschilds," 186–88, 195–96; Malone, *Dictionary*, 20:261–62; A. F. Williams, *Some Dreams*, 217–32; Davis, *English Rothschilds*, 212–15; Lockhart and Woodhouse, *Cecil Rhodes*, 103–5. The Exploration Company played a key role in the association between the English Rothschilds and Werner, Beit, and Company, a firm that had rapidly established control over gold production on the Central Rand. The Exploration Company also promoted and contributed to the success of deep-level mining, a method of extracting gold from mile-long underground veins of gold. The company expanded its scope in the 1890s. Hamilton Smith applied his expert knowledge of mining engineering to the problems of deep-level railway construction in London and Paris, which subsequently hired the Exploration Company to undertake these projects. Rhodes also involved Lord Nathan Rothschild in his most pressing projects, from the Rhodes Scholarships and the Cape-to-Cairo Railway project to the BSA. Rhodes even had Rothschild use his influence with the British government to prevent Colonial Office interference with Rhodes's projects in South Africa.

18 J. H. Hammond, "Gold Mines," 733.

19 Hall, "Rise of the Mining Industry," 9.

20 Ibid., 15–16.

21 Ibid., 11, 13–14.

22 J. H. Hammond, "Gold-Mining," 852–53.

23 Leroy-Beaulieu, "Labor Question," 256; Hall, "Rise of the Mining Industry," 18.

24 Wheatcroft, *Randlords*, 1–10.

25 "Native Labor," 47.

26 A. C. Sutherland, "Some Statistics of the Mining Industry of the Transvaal," MSS Afr. s. 1434, Rhodes House Library, Oxford University, 27–28.

27 "New Gold Fields," 373; Kubicek, *Economic Imperialism*, 51–52; Hall, "Rise of the Mining Industry," 18; Bailey, *Supplying*; Paul Johnson, *Consolidated Gold Fields*, 25–27.

28 J. H. Hammond, *Autobiography*, 1:201, 208, 213–14; *Engineering and Mining Journal*, January 18, 1896, 60.

29 J. H. Hammond, *Autobiography*, 1:197–212.

30 Henry D. Boyle to Cecil Rhodes, November 20, 1892, MSS Afr. s. 228, vol. C10, Rhodes Papers; "Hamilton Smith," July 14, 1900, 34; J. H. Hammond, *Autobiography*, 1:214.

31 Cartwright, *Gold*, 74–75; Kubicek, *Economic Imperialism*, 103.

32 Francis Rhodes to Cecil Rhodes, October 25, 1895, MSS Afr. s. 228, vol. C10 (2), Rhodes Papers.

33 "Henry C. Perkins, An Interview," in Rickard, *Interviews*, 413–29; Hall, *Mr. W. Ham. Hall's Statement*, 5–6.

34 William Hammond Hall to Hamilton Smith, April 8, 1891, box 5, folder 4, Hall Papers, Bancroft Library.

35 John Hays Hammond to Henry C. Perkins, October 7, 1896, John Hays Hammond to Secretary, December 2, 11, 21, 1896, December 9, 1897, John Hays Hammond to H. E. M. Davies, January 15, February 16, 1897, John Hays Hammond to William Hammond Hall, February 18, 1897, John Hays Hammond to Lord Harris, October 6, 1897, John Hays Hammond to Messrs. Wernher, Beit, and Company, October 15, 1897, Hammond private letterbook; John Hays Hammond to William Hammond Hall, October 16, November 17, 1896, Hammond private letterbook.

36 William Hammond Hall to John Hays Hammond, February 3, 15, 1897, box 5, folder 9, Hall Papers, Bancroft Library.

37 John Hays Hammond to William Hammond Hall, August 1, 1894, October 16, 1896, John Hays Hammond to Cecil Rhodes, February 18, 1897, John Hays Hammond to Lionel Phillips, February 18, 1897, Hammond private letterbook; William Hammond Hall to John Hays Hammond, February 3, 1897, box 5, folder 9, Hall Papers, Bancroft Library.

38 Agreement between Messrs. Wernher, Beit, and Company, the Consolidated Gold Fields of S.A. Limited, and William Hammond Hall, [1898], box 11, folder 57, Hall Papers, Bancroft Library.

39 William Hammond Hall, "General Report of Sept. 1897 to Messrs. W. B. & Co. and the Con. G.G. of S.A. Ltd. with Appendices," box 7, folder 24, Hall Papers, Bancroft Library; John Hays Hammond to Henry C. Perkins, October 7, 1896, John Hays Hammond to Secretary, December 2, 11, 21, 1896, December 9, 1897, John Hays Hammond to H. E. M. Davies, January 15, February 16, 1897, John Hays Hammond to William Hammond Hall, February 18, 1897, John Hays Hammond to Lord Harris, October 6, 1897, John Hays Hammond to Messrs. Wernher, Beit, and Company, October 15, 1897, Consolidated Gold Fields letterbook; John Hays Hammond to William Hammond Hall, October 16, November 17, 1896, Hammond private letterbook.

40 William Hammond Hall to Wernher, Beit, and Company and Directors of Consolidated Gold Fields of S.A. Ltd., July 4, 1898, box 6, folder 2, Hall Papers, Bancroft Library; William Hammond Hall to N. L. Spate, July 8, 1898, box 6, folder 2,
Hall Papers, Bancroft Library; William Hammond Hall to N. L. Spate, July 18,
1898, box 6, folder 1, Hall Papers, Bancroft Library.

41 Hall, "An American Engineer Abroad," box 6, folder 16, Hall Papers, Bancroft
Library.

42 Wernher, Beit, and Company to William Hammond Hall, January 20, 1899, box
17, folder 52, Hall Papers, Bancroft Library; William Hammond Hall, memorandum [1898?], box 6, folder 1, Hall Papers, Bancroft Library.

43 A. R. Goldring, "Notice of Meeting," April 6, 20, June 8, July 5, August 7, September 12, 1894, November 9, 1894, March 15, June 14, July 12, 1895, Witwatersrand
Chamber of Mines, Redwood Papers; "Grievances of the Uitlanders," February 5, 1897, box 37, folder 1, Central Mining Records, MSS British Empire, Rhodes
House Library, Oxford University, England; Wheatcroft, *Randlords*, 124–30;
"Transvaal Troubles," 82.

44 John Hays Hammond to John P. Jones, January 30, 1895, quoted in Spence, *Mining
Engineers*, 312.

45 Hole, *Jameson Raid*, 27.

46 Lowther, "Memoirs," 7.

47 I. Smith, *Origins*, 402; Paul Johnson, *Consolidated Gold Fields*, 32.

48 William Hammond Hall, "President S. J. P. Kruger," 23, 27, box 7, folder 5, Hall
Papers, Bancroft Library.

49 Layton, *Revolt*, 64–66.

50 Hall, *Mr. W. Ham. Hall's Statement*, 2, 5–6, 11–12.

51 J. H. Hammond, *Autobiography*, 1:335.

52 John Hays Hammond to Poultney Bigelow, August 28, 1896, Hammond private
letterbooks; N. H. Hammond, *Woman's Part*, 48.

53 Alan H. Jeeves, "The Consequences of the Raid," in Carruthers, *Jameson Raid*,
153–80.

54 Leander Starr Jameson to Gardner Williams, December 9, 1895, in "In the High
Court of Justice, Queen Bench's Division, *Regina v. Leander Starr Jameson*, Sir
John Christopher Willoughby, Bart., Henry Frederick White, Raleigh Grey,
Robert White, and Charles John Coventry, Offence under Foreign Enlistment
Act, 1870: Copies of Documents Put in as Exhibits to the Depositions at Bow
Street Police Court," 1896, MSS Afr. s 13/1, 13/2, Rhodes House Library, Oxford
University.

55 J. H. Hammond, *Autobiography*, 1:320–28; Blainey, "Lost Causes," 354–63; Spence,
*Mining Engineers*, 311–14; Hole, *Jameson Raid*, 25–30, 99–100; Shillington, *History*, 130–32; Macnab, *Gold*, 53–76; Wheatcroft, *Randlords*, 152–81. Contemporary accounts of the Jameson Raid include "Transvaal Crisis," 59–60; "Transvaal
Troubles," 81–82; "Dr. Jameson's Trial," *Standard*, March 11, 1896; "The Jameson
Trial," *Daily Graphic*, July 28, 1896; "The Transvaal Question: The Arrest of American Subject," *St. James Gazette*, January 14, 1896; "The Situation in South Africa,"
*Westminster Gazette*, January 16, 1896.

56 Macnab, *Gold*, 56.

57  Twain, *More Tramps Abroad*, 448–50.

58  J. H. Hammond, *Autobiography*, 1:218.

59  N. H. Hammond, *Woman's Part*, 48.

60  John Hays Hammond to Poultney Bigelow, August 28, 1896, Hammond private letter book.

61  J. H. Hammond, *Autobiography*, 1:4; Sackett, *Engineer*, 189–92.

62  Jacqueline Jaffe, "John Hays Hammond," in Carruthers, *Jameson Raid*, 211–26.

63  J. H. Hammond, *Autobiography*, 1:334, 343; N. H. Hammond, *Woman's Part*, 11; Macnab, *Gold*, 58.

64  Jaffe, "John Hays Hammond," 214.

65  Hall, "The South African Problem," 14, box 6, folder 1, Hall Papers, Bancroft Library; William Hammond Hall, "American Views on South Africa," 14, box 7, folder 1, Hall Papers, Bancroft Library.

66  Hall, "American Views," 7.

67  J. H. Hammond, *Truth*, 26–27.

68  Alfred Beit to Cecil Rhodes, October 2, 1897, vol. C24, Rhodes Papers.

69  J. H. Hammond, *Autobiography*, 2:408.

70  Shillington, *History*, 130–32.

71  Hall, "American Views," 4.

72  Hall, *Mr. W. Ham. Hall's Statement*, 5.

73  Bryce, *Impressions*, 432–33.

74  Beinart and Coates, *Environment and History*, 56–59.

75  William Hammond Hall to Samuel Marks, March 24, 1898, box 6, folder 1, Hall Papers, Bancroft Library; Hall, "South African Problem," 19–20; Cape of Good Hope, Department of Public Works, *Report upon Irrigation*, 44.

76  Bryce, *Impressions*, 439–40.

77  Hall, *Mr. W. Ham. Hall's Statement*, 4–6.

78  Ibid., 5–6; Cape of Good Hope, Department of Public Works, *Report upon Irrigation*, 1.

79  Hall, "Mr. President," box 6, folder 5, Hall Papers, Bancroft Library; William Hammond Hall to James Sivewright, March 14, 1899, box 6, folder 3, Hall Papers, Bancroft Library; "Conditions of the Transvaal and Rhodesia: An Entertaining Interview with Ex–State Engineer William Hammond Hall Regarding His Visit to South Africa," *San Francisco Chronicle*, November 7, 1897.

80  Weinmann, *Agricultural Research*, 1, 8.

81  William Hammond Hall to Katherine Buchanan Hall, August 4, 1897, box 5, folder 10, Hall Papers, Bancroft Library; "Conditions of the Transvaal."

82  William Hammond Hall, "Specifications for the Construction of an Earth-Work Dam for the Matoppo Reservoir, Sauerdale Farm near Bulawayo, Rhodesia," box 7, folder 18, Hall Papers, Bancroft Library.

83  William Hammond Hall, "Inkoos! Bahba! Mr. Rhodes and the Native Question," 1–2, 9–10, box 7, folder 3, Hall Papers, Bancroft Library.

84  William Hammond Hall to Katherine Buchanan Hall, August 4, 1897, box 5, folder 10, Hall Papers, Bancroft Library; "Conditions of the Transvaal."

85  Hall, "Inkoos! Bahba!" 8–9.

86  William Hammond Hall, "The Matabele Native Labor Question: Principles Which Operate at Home," 63, box 7, folder 9, Hall Papers, Bancroft Library.

87  Ibid., 62.

88  Ibid., 64–68.

89  William Hammond Hall, "Rt. Hon. Cecil John Rhodes," 34, box 7, folder 3, Hall Papers, Bancroft Library.

90  William Hammond Hall, "World Politics," 11, box 7, folder 3, Hall Papers, Bancroft Library.

91  Alfred Mosely to Cecil J. Rhodes, May 10, 1899, MSS Afr. s. 228, vol. 38, Rhodes Papers.

92  Herbert Baker to the Marquess of Lothian, April 2, 1937, H. E. V. Pickstone to Herbert Baker, March 5, 1937, Miscellaneous Papers Concerning C. J. Rhodes, MSS Afr. s. 8, Rhodes House Library, Oxford University; "Advances Made by the Rt. Hon. C. J. Rhodes, Mr. A. Beit and the De Beers Mines in Respect of the Rhodes Fruit Farms and Other Ventures, Cape Town, Aug. 13, 1901," MSS Afr. s. 228, vol. C8, Rhodes Papers; Lockhart and Woodhouse, *Cecil Rhodes*, 201–3; H. E. V. Pickstone, memorandum, n.d., MSS Afr. s. 228, vol. C8, Rhodes Papers.

93  H. E. V. Pickstone to S. J. du Toit, September 19, 1901, MSS Afr. s. 228, vol. C8, Rhodes Papers; H. E. V. Pickstone, "2nd Meeting of Managers at 'Lekkerwyn,' Wednesday, August 23, 1899," MSS Afr. s. 228, vol. C8, Rhodes Papers; Alfred Mosely to Cecil J. Rhodes, May 3, 1899, MSS Afr. s. 228, vol. C8, Rhodes Papers; H. E. V. Pickstone to Cecil J. Rhodes, August 7, 1901, MSS Afr. s. 228, vol. C8, Rhodes Papers; Pickstone, *Report*, 1; Alfred Mosely to Cecil J. Rhodes, May 10, 1899, MSS Afr. s. 228, vol. C8, Rhodes Papers.

94  Unidentified to Cecil Rhodes, November 25, 1901, Rhodes Papers, MSS Afr. t5, Rhodes House Library, Oxford University.

95  William Hammond Hall to Katherine Hall, July 5, 1897, box 5, folder 10, Hall Papers, Bancroft Library; John Hays Hammond to Cecil J. Rhodes, February 18, 1897, Hammond private letterbook.

96  "South Africa," in U.S. Senate, *Report*, 308–12; Cape of Good Hope, Department of Public Works, *Report upon the Grobbelaar's River*; Cape of Good Hope, *Douglas Irrigation Works*; Cape of Good Hope, *Report of the Thebus Irrigation Commission*.

97  James Sivewright to William Hammond Hall, April 21, 1897, September 5, 1898, November 18, 1898, box 16, folder 16, Hall Papers, Bancroft Library.

98  William Hammond Hall to James Sivewright, March 14, 1899, box 6, folder 3, Hall Papers, Bancroft Library; "Conditions of the Transvaal"; William Hammond Hall to editor of *South Africa*, [1898], box 6, folder 1, Hall Papers, Bancroft Library.

99  Hall, "Rise of the Mining Industry," 19–20.

100  Alfred Mosely to Cecil J. Rhodes, May 10, 1899, MSS Afr. s. 228, vol. 38, Rhodes Papers.

101  Bryce, *Impressions*, 397.

102  William Hammond Hall to editor of *South Africa*, [1898], box 6, folder 1, Hall Papers, Bancroft Library.

103  J. N. Smartt to William Hammond Hall, July 18, 1900, box 17, folder 18, Hall Papers, Bancroft Library.

104 William Hammond Hall to James Sivewright, May 12, 1897, box 5, folder 9, Hall Papers, Bancroft Library.

105 William Hammond Hall to editor of *South Africa*, [1898], box 6, folder 1, Hall Papers, Bancroft Library.

106 Cape of Good Hope, *Report on Irrigation Legislation*, 64–69; Cape of Good Hope, Department of Public Works, *Report upon Irrigation*, 4.

107 William Hammond Hall to Samuel Marks, March 24, 1898, box 6, folder 1, Hall Papers, Bancroft Library.

108 Cape of Good Hope, *Report on Irrigation Legislation*, 62.

109 William Hammond Hall to Samuel Marks, March 24, 1898, box 6, folder 1, Hall Papers, Bancroft Library.

110 Hall, "South African Problem," 19–20.

111 "Agricultural Possibilities," 344.

112 Cape of Good Hope, *Report on Irrigation Legislation*, 59–69; Hall, "Rise of the Mining Industry," 24; "Conditions of the Transvaal."

113 William Hammond Hall, "Boers and Burghers (2)," 3, 18, 11, box 6, folder 17, Hall Papers, Bancroft Library.

114 Hall, "Mr. President," 24; William Hammond Hall, "Unwise Impositions: Race Feeling Becoming Dominant," 9–20, box 6, folder 1, Hall Papers, Bancroft Library; William Hammond Hall, "After the War Is Over: The Regeneration of South Africa," 9–20, Hall Papers, Bancroft Library.

115 Shaun Milton, "The Transvaal Beef Frontier: Environment, Markets, and the Ideology of Development, 1902–1942," in *Ecology and Empire*, ed. Griffiths and Robin, 199–212; Lowther, "Memoirs," 57.

116 Leroy-Beaulieu, "Labor Question," 256; "Labor Conditions in Johannesburg," 44; "Labor Conditions in the Transvaal Mines," 40; "Shipping Chinamen," 255; Carter, "Chinese"; Paul Johnson, *Consolidated Gold Fields*, 39–40.

### Chapter Five

1 Adler, *Claus Spreckels*, 3; Adler, "Maui Land Deal," 155.

2 Newell, *Hawaii*, 8–16; Cox, "Water Development," 175–80; L. H. Herschler, "Irrigation in the Hawaiian Islands," 1–9, Water Resources Center Archives, University of California, Berkeley; O'Shaughnessy, "Irrigation Works," 129–30.

3 Adler, *Claus Spreckels*, 16–17.

4 See Stevens, *American Expansion*.

5 Twain, *Letters*, 69.

6 *California Illustrated: Hawaiian Edition* (1897–98), 3.

7 Cordray, "Claus Spreckels," 2–6.

8 Ibid., 9–16; "The California Sugar Refinery, Its Founder, and His Work," *San Francisco Merchant*, May 27, 1881.

9 "California Sugar Refinery," 9.

10 *Something about Sugar*, 13.

11 Ibid., 6, 9; W. P. Alexander, *Irrigation*, 1; "Sugar-Refining, the Early History of Sugar-Making," *San Francisco Commercial Herald*, January 15, 1880.

12 Adler, *Claus Spreckels*, 12; Wilcox, *Sugar Water*, 12–13.

13  *California Illustrated: Hawaii Edition* (1897–98), 114.

14  Cordray, "Claus Spreckels," 32–36; Adler, *Claus Spreckels*, 3.

15  "Spring and the Grocery Trade as Viewed by Leading Businessmen," *San Francisco Weekly Journal of Commerce*, March 10, 1881.

16  Adler, "Maui Land Deal," 155–56.

17  "The Sugar Industry of the Hawaiian Islands," *Honolulu Evening Bulletin*, industrial ed., November 1901, 12; Wilcox, *Sugar Water*, 57, 114–16, 122–23; *Story of Sugar*, 16–17; Gutleben, *Sugar Tramp*, 5; Vandercook, *King Cane*, 66–69.

18  Nordhoff, *Northern California*, 57.

19  Cordray, "Claus Spreckels," 36; *Claus Spreckels, Plaintiff, v. Hawaiian Commercial and Sugar Company, in the Superior Court of the State of California, in and for the City and County of San Francisco*, no. 53,045, p. 2; Wilcox, *Sugar Water*, 56–58; Adler, "Maui Land Deal," 163; Adler, *Claus Spreckels*, 16–19.

20  Newell, *Hawaii*, 25; Wadsworth, "Historical Summary," 131–32.

21  Cordray, "Claus Spreckels," 37–38.

22  Quoted in Adler, *Claus Spreckels*, 44.

23  Gutleben, *Sugar Tramp*, 8.

24  Adler, *Claus Spreckels*, 27.

25  *Claus Spreckels, Plaintiff, v. Hawaiian Commercial and Sugar Company*, 2–3; Adler, *Claus Spreckels*, 12–13.

26  Quoted in Cordray, "Claus Spreckels," 42.

27  Wilcox, *Sugar Water*, 16.

28  Adler, *Claus Spreckels*, 46; *Claus Spreckels, Plaintiff, v. Hawaiian Commercial and Sugar Company*, 2.

29  O'Shaughnessy, "Irrigation Works," 134; Vandercook, *King Cane*, 69–70; Cordray, "Claus Spreckels," 36–39; B. Adams, "Sugar Irrigation," 125; Adler, *Claus Spreckels*, 49.

30  "Sugar-Refining."

31  O'Shaughnessy, "Irrigation Works," 137.

32  Adler, *Claus Spreckels*, 72.

33  *Story of Sugar*, 58–59.

34  Cordray, "Claus Spreckels," 39–40; Adler, *Claus Spreckels*, 70–71, 77; *Story of Sugar*, 60–63.

35  Condé and Best, *Sugar Trains*, 208–9; "Sugar Industry," 12; Adler, *Claus Spreckels*, 70–71.

36  Claus Spreckels, "Pacific Steamship Line," Bancroft Library, University of California, Berkeley; Spreckels and Brothers, *Ports*, 47; Adler, *Claus Spreckels*, 105.

37  Thurston, *Labor Situation*, 9.

38  Adler, *Claus Spreckels*, 6.

39  Nordhoff, *Northern California*, 63.

40  Twain, *Letters*, 51–52.

41  Baldwin, "Sugar Industry," 668; Marutani, "Labor-Management Relations," 11–13, 19–20; Adler, *Claus Spreckels*, 233.

42  *Story of Sugar*, 25–32; Adler, *Claus Spreckels*, 7.

43  Adler, *Claus Spreckels*, 77.

44  Twain, *Letters*, 53.

45 Baldwin, "Sugar Industry," 668.
46 Thurston, *Labor Situation*, 6–7.
47 Baldwin, "Sugar Industry," 668.
48 Twain, *Letters*, 38.
49 Thurston, *Labor Situation*, 13–14, 16.
50 Newell, *Hawaii*, 42.
51 Quoted in Adler, *Claus Spreckels*, 231.
52 Cordray, "Claus Spreckels," 63; Marutani, "Labor-Management Relations," 11–20.
53 Adler, *Claus Spreckels*, 233, 249.
54 Newell, *Hawaii*, 46–47.
55 "Sugar-Refining."
56 Newell, *Hawaii*, 6.
57 Ibid., 49.
58 Ibid., 45.
59 Cordray, "Claus Spreckels," 51–52.
60 "Spring and the Grocery Trade"; "California Sugar Refinery"; "The Laying of the Corner Stone," *Daily Alta San Francisco*, May 29, 1881; "California Sugar Refinery," 1, Bancroft Library, University of California, Berkeley.
61 "California Sugar Refinery."
62 Sager, "From Memory's Files," 7, 137; "California Sugar Refinery."
63 Cordray, "Claus Spreckels," 71–75.
64 Wilcox, *Sugar Water*, 114, 122; Condé and Best, *Sugar Trains*, 210; "Sugar Industry," 13; *Story of Sugar*, 94.
65 Schuyler, *Report*; Schuyler, *Culture*; James D. Schuyler, "Report on Water Development for Oahu Plantation," April 29, 1907, Schuyler 182, Water Resources Center Archives, University of California, Berkeley; James D. Schuyler, "Report on the Condition and Probable Future Safety of the Uncompleted Nuuanu Dam," April 20, 1907, Schuyler 183.3, Water Resources Center Archives, University of California, Berkeley; Kieffer, Grunsky, and Lippincott, "Memoir of James Dix Schuyler," 2243–45.
66 Joseph B. Lippincott, "Report on the Feasibility of Brining the Waters of the Waiahole, Waikane, and Kahana Streams through the Koolau Range to the Lands of the Oahu Sugar Company, Limited," August 19, 1911, Lippincott 42.1, Lippincott Papers; letters between Lippincott and the Oahu Sugar Company, July 6, 1911–January 6, 1912, Lippincott 42.2, Lippincott Papers.
67 O'Shaughnessy, "Irrigation Works," 129–37.
68 George P. Cooke to Elwood Mead, December 7, 1921, box 1, correspondence between Elwood Mead and Hawaiian Homes Commission, Mead Papers, Bancroft Library.
69 Kauanui, "Rehabilitating," 3, 5.
70 Cooke, *Moolelo O Molokai*, 76; Vause, "Hawaiian Homes Commission Act," 1.
71 Vause, "Hawaiian Homes Commission Act," 73, 116–17; Kluger, *Turning on Water*, 103–4.
72 Kauanui, "Rehabilitating," 2–6, 46; Kluger, *Turning on Water*, 103–4; Vause, "Hawaiian Homes Commission Act," iv.
73 Bishop, *Why Are the Hawaiians Dying Out?*, 3–11, 18.

74 Kauanui, "Rehabilitating," 85, 102; Alan Murakami, "The Hawaiian Homes Commission Act," in *Native Hawaiian Rights Handbook*, ed. MacKenzie, 44.

75 Kauanui, "Rehabilitating," 86, 104; "Hawaiian Board Is Advised by Mead," *Berkeley (California) Gazette*, July 26, 1922; Vause, "Hawaiian Homes Commission Act," 24–26.

76 Hawaiian Homes Commission, *Rehabilitation in Hawaii*, 18, 3.

77 Ibid., 11–15.

78 Quoted in Kluger, *Turning on Water*, 104.

79 Keesing, *Hawaiian Homesteading*, 25, 27. Population figures for Molokai do not include the leper colony, Kalaupapa, formed in 1864 in north-central Molokai.

80 Ibid., 27–28, 67.

81 Hawaiian Homes Commission, *Rehabilitation in Hawaii*, 21.

82 Cooke, *Moolelo O Molokai*, 77; "Eight Chosen for Places on Molokai Land," *Honolulu Star-Bulletin*, June 28, 1922.

83 Elwood Mead to Hawaiian Homes Commission, January 12, 1923, box 5, Mead Papers, Bancroft Library.

84 "Some Facts about the Haiku Homestead District, Haiku, Maui Co., Hawaii," box 5, Mead Papers, Bancroft Library.

85 Charles J. McCarthy to John B. Payne, June 30, 1920, quoted in Vause, "Hawaiian Homes Commission Act," 117.

86 "Dr. Mead Airs His Views on Colonization," *Honolulu Star-Bulletin*, June 28, 1922.

87 William Atherton DuPuy quoted in Kauanui, "Rehabilitating," 66.

88 "Ex-Soldiers to Colonize Rural Areas in Hawaii," *Los Angeles Herald*, September 8, 1922.

89 Elwood Mead to Hawaiian Homes Commission, January 12, 1923, box 5, Mead Papers, Bancroft Library.

90 Newell, *Hawaii*, 7.

91 Elwood Mead to Hawaiian Homes Commission, January 12, 1923, box 5, Mead Papers, Bancroft Library.

92 Keesing, *Hawaiian Homesteading*, 62.

93 Elwood Mead to Hawaiian Homes Commission, January 12, 1923, box 5, Mead Papers, Bancroft Library.

94 Ibid.; Kluger, *Turning on Water*, 104.

95 Elwood Mead to Hawaiian Homes Commission, January 12, 1923, box 5, Mead Papers, Bancroft Library.

96 "Record of Visit to Waiakea Homesteads," June 22, 1922, 2, 3, item 2, Hawaiian Homes Commission I, Mead Papers, Bancroft Library.

97 Elwood Mead communication, October 20, 1927, in *Report of the Hawaiian Homes Commission*, 51.

98 "Record of Visit to Wailuku," Hawaiian Homes Commission I, Mead Papers, Bancroft Library, 2–5.

99 "Dr. Mead Says Hawaii's Colonization Plan Has Every Chance of Success," *Sacramento Bee*, July 17, 1922; W. R. Farrington to Elwood Mead, July 6, 1922, box 1, Mead Papers, Bancroft Library.

100 Keesing, *Hawaiian Homesteading*, 45–49.

101 Ibid., 56, 63–64, 65.

102 Ibid., 110, 123.

103 Kluger, *Turning on Water*, 104–5.

104 Keesing, *Hawaiian Homesteading*, 91–92.

105 Thurston, *Labor Situation*, 3–4.

106 Elwood Mead to George P. Cook, January 17, 1922, box 1, Mead Papers, Bancroft Library.

107 Du Puy, *Hawaii and Its Race Problem*, 100.

108 "Notes of the Trip to Hawaii, Written in the Union Club, Honolulu, June 29, 1922," 1, Hawaiian Homes Commission Report, 1922, Mead Papers, Bancroft Library.

109 *California Illustrated: Hawaiian Edition* (1897–98), 3.

110 "Notes of the Trip to Hawaii," 1.

111 Du Puy, *Hawaii and Its Race Problem*, 115–17.

112 Keesing, *Hawaiian Homesteading*, 121, 130.

113 Ibid., 27–28, 67.

114 Elwood Mead communication, October 20, 1927, in *Report of the Hawaiian Homes Commission*, 51.

115 Ibid., 13.


## Chapter Six

1 Kluger, *Turning on Water*, 102–14.

2 See Troen, "Frontier Myths."

3 Kluger, *Turning on Water*, 107.

4 Ruppin, *Agricultural Colonisation*, 171; Pearlman, *Collective Adventure*, 75; Mead et al., *Agricultural Colonization*, 13; Mead, "New Palestine," 623–29; Kluger, *Turning on Water*, 107.

5 Bein, *History*, 15.

6 Morris, *Righteous Victims*, 37.

7 Revusky, *Jews*, 5–8; Bein, *History*, 3–15; Aaronsohn, "Beginnings," 438. Bilu is an acronym for "Beth Ya'akov lekhu venelkha" (O house of Jacob, come ye, and let us walk [Isaiah 2:5]).

8 *Report of the High Commissioner on the Administration of Palestine, 1920–25*, quoted in Horowitz, *Jewish Colonisation*, 26.

9 Twain, *Innocents Abroad*, 465–66.

10 Mead et al., *Agricultural Colonization*, 17; Bein, *History*, 6–15; Revusky, *Jews*, 57–58; *Area of Cultivable Land*, 2.

11 Frank Adams, "Palestine Agriculture in the Light of Recent Investigations," [193?], 4–5, 11–12, 15–16, Frank Adams Papers, Adams 137; Bein, *History*, 6–15; Revusky, *Jews*, 57–58; *Area of Cultivable Land*, 2; Morris, *Righteous Victims*, 3–5.

12 Mead et al., *Agricultural Colonization*, 34; Bein, *Return*, 6–7.

13 Ussishkin, "Jewish Colonisation Association"; Bein, *History*, 8–15; Norman, *Outstretched Arm*, 55–61.

14 Bein, *History*, 8–15; Revusky, *Jews*, 11; Norman, *Outstretched Arm*, 58–61.

15 Herzl, *Jewish State*, 14.

16 Penslar, "Engineering Utopia," 64; Bein, *History*, 16–26; Revusky, *Jews*, 13–27,

137–42; Schorske, *Fin-de-Siècle Vienna*, 117–75; Bein, *Return*, 11–17; Vital, *Origins*, 233–66.

17  Ruppin, *Agricultural Colonisation*, 129–44; Bein, *History*, 16–26; Revusky, *Jews*, 13–27, 137–42.

18  Pearlman, *Collective Adventure*, 41, 25–26.

19  Quoted in Horowitz, *Jewish Colonisation*, 26.

20  Morris, *Righteous Victims*, 43; Oettinger, *Jewish Colonization*, 2–30, 92–108; Aaronsohn, "Beginnings," 443–44; Bein, *History*, 15; Kamen, *Little Common Ground*, 11–12.

21  Mead, *Agricultural Development*, 9; Revusky, *Jews*, 21–27, 125–47; Kluger, *Turning on Water*, 108.

22  Mead, *Agricultural Development*, 2; Frank Adams, "Outline for Talk at B'nai Brith Hall, San Francisco, April 30, 1928: Observations in the Zionist Colonies in Palestine," 1, Adams Papers, Adams 144; Kluger, *Turning on Water*, 108.

23  Elwood Mead to Adolf Koshland, February 17, 11, 1922, January 20, 1920, box 1, Mead Papers, Bancroft Library; S. E. H. Soskin to Elwood Mead, December 23, 1921, March 21, 1922, Elwood Mead to S. E. H. Soskin, February 23, 1922, box 2, Mead Papers, Bancroft Library; Adolf Koshland to Elwood Mead, January 28, 1922, J. M. Golden to Elwood Mead, February 17, 1922, box 1, Mead Papers, Bancroft Library; Ruppin, *Agricultural Colonisation*, 171; Kluger, *Turning on Water*, 108.

24  Mead, "New Palestine," 124.

25  Mead, *Agricultural Development*, 2–6; Karlinsky, "California Dreaming," 24–40; Kluger, *Turning on Water*, 108–9.

26  Mead et al., *Agricultural Colonization*, 18, 20, 23–26; F. Adams, "Frank Adams," 355; F. Adams, "Agriculture," 15; Revusky, *Jews*, 57.

27  Pearlman, *Collective Adventure*, 164.

28  Mead, "Farming," 416; Mead, *Agricultural Development*, 7–8.

29  Leo Wolman, "The Labor Situation in Palestine," in Joint Palestine Survey Commission et al., *Reports*, 511; Kluger, *Turning on Water*, 109.

30  Kluger, *Turning on Water*, 109–10; Mead, *Agricultural Development*, 15–24.

31  Mead, "New Palestine," 624.

32  Ibid., 628.

33  Mead, *Agricultural Development*, 2–6, 24.

34  Ibid., 11; Ryerson, *World*, 147; F. Adams, "Frank Adams," 352–53, 359–60.

35  Mead et al., *Agricultural Colonization*, 37, 18; Kluger, *Turning on Water*, 108.

36  Mead et al., *Agricultural Colonization*, 13.

37  F. Adams, "Outline," 3–4.

38  A. T. Strahorn, "Soil Reconnaissance," in Joint Palestine Survey Commission et al., *Reports*, 223.

39  Knowles A. Ryerson, "The Horticultural Possibilities as Especially Related to Agricultural Colonization," in ibid., 379–81.

40  Ibid., 382; Ryerson, *World*, 150–53.

41  Jacob G. Lipman, "Report on Education and Research," in Joint Palestine Survey Commission et al., *Reports*, 422, 423.

42  Ibid., 425–30.

43 Ryerson, "Horticultural Possibilities," 257.

44 Mead et al., *Agricultural Colonization*, 37.

45 Ryerson, "Horticultural Possibilities," 294.

46 Mead et al., *Agricultural Colonization*, 31.

47 Ryerson, "Horticultural Possibilities," 275, 290.

48 Ibid., 296–303; Wolman, "Labor Situation," 501; F. Adams, "Agriculture," 15.

49 F. Adams, "Agriculture," 28, 26.

50 Mead et al., *Agricultural Colonization*, 41.

51 Ibid., 38; Kluger, *Turning on Water*, 110.

52 Pearlman, *Collective Adventure*, 73.

53 Ibid., 74.

54 Sir John Campbell, "Report on the Jewish Settlements," Joint Palestine Survey Commission et al., *Reports*, 435, 436.

55 Ruppin, *Agricultural Colonisation*, 176.

56 Campbell, "Report," 436.

57 Kluger, *Turning on Water*, 110.

58 Campbell, "Report,"437–38, 475.

59 F. Adams, "Frank Adams," 356–57; Ruppin, *Agricultural Colonisation*, 137.

60 Mead et al., *Agricultural Colonization*, 35.

61 Ibid., 60–63; Frank Adams, "Outline of Talk on 'Some Agricultural Observations in Palestine': Before Staff Seminar at Berkeley," January 25, 1928, Adams Papers, Adams 144.

62 Mead et al., *Agricultural Colonization*, 43.

63 Leo Wolman, "The Labor Movement and Its Activities," in Joint Palestine Survey Commission et al., *Reports*, 515–35.

64 Ryerson, "Horticultural Possibilities," 245.

65 Mead et al., *Agricultural Colonization*, 40–58.

66 See Lowdermilk, *Palestine*; J. B. Hays, *TVA*; Rook, "American in Palestine."

67 Rook, "American in Palestine."

68 Ibid.


## Conclusion

1 Newell, "Awakening," 568.

2 Morison, "Address."

3 Wolf, *Europe*, 391.

4 Richard Cobden quoted in Porter, *Lion's Share*, 6.

5 Headrick, *Tools*, 205; Porter, *Lion's Share*, 68–89.

6 Headrick, *Tentacles*, 3–9, 104–10.

7 Headrick, *Tools*, 205–6.

8 J. H. Hammond, *Engineer*, 9.

9 Headrick, *Tools*, 208.

10 J. H. Hammond, *Engineer*, 10.

11 Layton, *Revolt*, 16, 29.

12 J. H. Hammond, *Engineer*, 16.

13 Ibid., 120.

14  Veblen, *Engineers*, 74; Bledstein, *Culture*, 1–45, Layton, *Revolt*, 64.
15  Mumford, *Technics*, 219–20.
16  J. H. Hammond, *Engineer*, 192–93.
17  Layton, *Revolt*, 189–91.
18  Ibid., 189–90, 231, 203–9; see also Rickard et al., *Economics*; Clements, *Hoover*; George H. Nash, *Life*.
19  Pisani, *From the Family Farm*, 154–90.
20  Layton, *Revolt*, 64.
21  Ibid., 29.
22  Ibid., x.
23  Ibid., 66.
24  Newell quoted in ibid., 118.
25  J. H. Hammond, *Engineer*, 29.
26  Layton, *Revolt*, x; William M. Evan, "The Engineering Profession: A Cross Cultural Analysis," in *Engineers*, ed. Perrucci and Gerstl, 117–25.
27  J. H. Hammond, *Engineer*, 37.

## Archival Sources

Adams, Frank. "Frank Adams, University of California, on Irrigation, Reclamation, and Water Administration." Interview by Willa Klug Baum. 1959. Regional Oral History Office, Bancroft Library, University of California, Berkeley.

———. Papers, 1899–1962. Water Resources Center Archives, University of California, Berkeley.

British South Africa Company. Reports, Minutes, Papers, and Correspondence, 1888–1911. Rhodes House Library, Oxford University, England.

Consolidated Gold Fields of South Africa Ltd. Letterbook, July 1896–December 1897. Rhodes House Library, Oxford University, England.

Culcheth, W. W. "Irrigation in India: Its Lessons for Australia: A Paper Read by Mr. W. W. Culcheth before the Victorian Institute of Engineers, June 3, 1891." State Library of Victoria, Melbourne, Australia.

Davidson, George. Papers, 1825–1911. Bancroft Library, University of California, Berkeley.

Deakin, Alfred. "Irrigation in Australia [1891, 1901]." National Library of Australia, Canberra.

———. Notebooks, 1905–14. National Library of Australia, Canberra.

———. Papers, 1804–1973. National Library of Australia, Canberra.

Gutleben, Dan. Papers. Bancroft Library, University of California, Berkeley.

Hall, William Hammond. "Irrigation in California." Lecture Delivered before the National Geographic Society in Washington, January 25, 1889. Washington, D.C.

———. Papers, 1803–1979. Bancroft Library, University of California, Berkeley.

———. Papers, 1873–1911. California Historical Society, San Francisco.

Hammond, John Hays. Private Letterbooks, 1893–1907. 4 vols. Rhodes House Library, Oxford University, England.

Lippincott, Joseph B. Papers. Water Resources Center Archives, University of California, Berkeley.

Lowther, Muriel. "Memoirs of George Farrar: Mining in South Africa, Jameson Raid, &c., 1893–1952." Rhodes House Library, Oxford University, England.

Mead, Elwood. Correspondence, 1912–21. Water Resources Center Archives, University of California, Berkeley.

———. "The Distribution of Water in Irrigation in Australia: A Paper to Be Presented at a Meeting of the International Engineering Congress, 1915, in San Francisco, California." National Library of Australia, Canberra.

————. Papers, 1900–1942. Water Resources Center Archives, University of California, Berkeley.

————. Papers, 1907–24. Bancroft Library, University of California, Berkeley.

————. "Rural Credits in Australia." University of California, Berkeley.

Memorandum of Association of the Exploration Company, Limited, October 26, 1886. Public Record Office, London, sec. 148, ref. BT 31 3750 23378.

Memorandum of Association of the Exploration Company, Limited, October 21, 1889. Public Record Office, London, sec. 148, ref. BT 31 4570 29924.

Panama Pacific International Exposition Pamphlets. Bancroft Library, University of California, Berkeley.

Quinan, William Russell. Autobiography and Papers, 1877–1923. 6 vols. Rhodes House Library, Oxford University, England.

Register of Managers of the Exploration Company, Limited, November 27, 1886. Public Record Office, London, sec. 148, ref. BT 31 4570 29924.

Rhodes, Cecil J. British South Africa Company Papers, 1890–1903. Rhodes House Library, Oxford University, England.

————. Papers. Rhodes House Library, Oxford University, England.

————. Scrapbook, 1896–1902. Rhodes House Library, Oxford University, England.

Rhodes Fruit Farms Ltd. Photograph Albums. 2 vols. Rhodes House Library, Oxford University, England.

Rural Water Corporation Collection. State Library of Victoria, Australia.

Sager, Henry. "From Memory's Files: Episodes in the Life of a Western Sugar Refinery Employee." Bancroft Library, University of California, Berkeley.

Schuyler, James Dix. Papers. Water Resources Center Archives, University of California, Berkeley.

Wescoat, James L., Jr. "Water Rights in South Asia and the United States: Comparative Perspectives, 1873–1998." Comparative Examination Landed Property Rights Project, Social Science Research Council, January 20, 1991.

Witwatersrand Chamber of Mines. Charles Lewis Redwood Papers, 1886–1948. Rhodes House Library, Oxford University, England.

Works, John D. Papers, 1910–17. Bancroft Library, University of California, Berkeley.


### Books, Essays, Articles, and Dissertations

Aaronsohn, Ran. "The Beginnings of Modern Jewish Agriculture in Palestine: 'Indigenous' versus 'Imported.'" *Agricultural History* 69, no. 3 (1995): 438–53.

"Abstracts of Official Reports." *Engineering and Mining Journal*, April 24, 1897.

Adams, Bristow. "Sugar Irrigation in Hawaii." *Forestry and Irrigation*, 9, no. 3 (January 1903): 122–31.

Adams, Frank. "Agriculture in Palestine." *California Countryman* 14, no. 5 (1928): 15–21.

————. "Data on Land Settlement in California." *Transactions of the Commonwealth Club of California* 11, no. 8 (1916): 375–96.

————. "Land Settlement." *Transactions of the Commonwealth Club of California* 10, no. 5 (1915): 369–428.

Adams, Mildred. "Native Values." *Woman Citizen*, January 14, 1922, 9.

Adler, Jacob. *Claus Spreckels: The Sugar King in Hawaii*. Honolulu: University of Hawaii Press, 1966.

———. "The Maui Land Deal: A Chapter in Claus Spreckels' Hawaiian Career." *Agricultural History* 39, no. 3 (1965): 155–63.

*Agreement between the Government of the Colony of Victoria and George Chaffey and William Benjamin Chaffey to Secure the Application of Private Capital to the Construction of Irrigation Works* . . . Melbourne: Government Printer, 1887.

"Agricultural Possibilities of the Transvaal: Views of the Australian." *Irrigation Age* 14, no. 10 (July 1900): 342–45.

Alexander, J. A. *The Life of George Chaffey: A Story of Irrigation Beginnings in California and Australia*. Melbourne: Macmillan, 1928.

Alexander, W. P. *The Irrigation of Sugar Cane in Hawaii*. Honolulu: Experiment Station of the Hawaiian Sugar Planters' Association, 1923.

Ali, Imran. *The Punjab under Imperialism, 1885–1947*. Princeton: Princeton University Press, 1988.

Andrews, John. "Irrigation in Eastern Australia." *Australian Geographer* 3, no. 6 (1939): 14–15.

Anstey, Vera. *The Economic Development of India*. New York: Longman, 1929.

Aplin, Graeme, S. G. Foster, and Michael McKernan, eds. *Australians: A Historical Dictionary*. New South Wales: Fairfax, Syme, and Weldon, 1987.

*The Area of Cultivable Land in Palestine*. Jerusalem: Jewish Agency for Palestine, 1936.

Arnold, David, and Ramachandra Guha, eds. *Nature, Culture, Imperialism: Essays on the Environmental History of South Asia*. Delhi: Oxford University Press, 1995.

Aubury, Lewis E. "California Irrigation Needs Forests." *Forestry and Irrigation* 14, no. 4 (1908): 212–14.

"Australian Gold." *Mining and Scientific Press*, March 9, 1895.

"Australian Gold Production." *Engineering and Mining Journal*, March 1, 1902.

*The Australian Irrigation Colonies: Mildura, Renmark: Official Reports of the Victorian Water Supply Department and Reprinted Articles from the Leading Australian Journals*. Melbourne: Chaffey, 1891.

Awasthi, Aruna. *History and Development of Railways in India*. New Delhi: Deep and Deep, 1994.

Bailey, Lynn R. *Supplying the Mining World: The Mining Equipment Manufacturers of San Francisco, 1850–1900*. Tucson: Westernlore, 1996.

Baker, I. G. "Elwood Mead in Australia." *Aqua* 2, no. 6 (February 1951): 3–11.

Baldwin, H. P. "The Sugar Industry in Hawaii." *Overland Monthly* 25, no. 150 (June 1895): 663–68.

Ballantyne, James. *Our Colony in 1880: Pictorial and Descriptive*. Melbourne: Hutchinson, Glasgow, 1880.

Barth, Gunther. *Instant Cities: Urbanization and the Rise of San Francisco and Denver*. New York: Oxford University Press, 1975.

Baxter, William M., Jr. *Panama Canal*. N.p.: Isthmian Canal Commission, 1913.

Behr, E. C. "Colonization Methods Past and Present." *California Countryman* 8, no. 3 (1922).

Bein, Alex. *The History of Jewish Agricultural Settlement in Palestine*. 2nd ed. Jerusalem: Zionist Organisation Youth Department, 1945.

——. *The Return to the Soil: A History of Jewish Settlement in Israel.* Jerusalem: Zionist Organisation, 1952.

Beinart, William, and Peter Coates. *Environment and History: The Taming of Nature in the USA and South Africa.* London: Routledge, 1995.

Bellasis, E. S. *Punjab Rivers and Works: A Description of the Shifting Rivers of the Punjab Plains and of Works on Them . . .* Allahabad: Pioneer, 1911.

Benedict, Burton. *The Anthropology of World's Fairs: San Francisco's Panama Pacific International Exposition of 1915.* London: Lowie Museum of Anthropology, 1983.

Bennett, Claude N. "The Nation's Herculean Task." *Conservation* 14, no. 9 (1908): 506–8.

Bennett, Ira, John Hays Hammond, and Patrick J. Lennox. *History of the Panama Canal: Its Construction and Builders.* Washington, D.C.: Historical, 1915.

Bishop, S. E. *Why Are the Hawaiians Dying Out? or, Elements of Disability for Survival among the Hawaiian People.* Honolulu: n.p., 1888.

Blainey, Geoffrey. "Lost Causes of the Jameson Raid." *Economic History Review* 18, no. 2 (1965): 350–66.

——. *The Rush That Never Ended: A History of Australian Mining.* Melbourne: Melbourne University Press, 1963.

Bledstein, Burton J. *The Culture of Professionalism: The Middle Class and the Development of Higher Education in America.* New York: Norton, 1976.

Booth-Tucker, Commander. "The Relation of Colonization to Irrigation." *Forestry and Irrigation* 9, no. 10 (1903): 499–505.

Branagan, P. J. *The Railway Policy for Victoria: How to Assist the Agricultural Producer . . .* Melbourne: Fraser and Jenkinson, [1911?].

Brechin, Gray. *Imperial San Francisco: Urban Power, Earthly Ruin.* Berkeley: University of California Press, 1999.

Brereton, Robert M. *Extracts from a Report by R. M. Brereton, C.E., to Messrs. Friedlander, Ralston, and Chapman.* San Francisco: n.p., 1871.

——. *Reminiscences of an Old English Civil Engineer, 1858–1908.* Portland, Ore.: Irwin-Hodson, 1908.

——. *Report on Messrs. Bensley & Co.'s Canal Project from Fresno Slough to Tide Water at Antioch.* San Francisco: Bancroft, 1872.

Brown, H. "Irrigation under British Engineers." *Transactions of the American Society of Civil Engineers* 54, pt. C (1904): 3–31.

Bryce, James. *Impressions of South Africa.* London: Macmillan, 1900.

Buckley, Robert Burton. *The Irrigation Works of India.* 2nd ed. London: Spon, 1905.

Bunje, E. T. H., F. J. Schmitz, and D. T. Wainwright. *Careers of University of California Mining Engineers, 1865–1936.* Berkeley: California Cultural Research Project, 1936.

Burton, A. R. E. *How to Get Rich in Victoria: Mildura, the True Australia Felix.* Melbourne: Spectator, 1892.

California. Commission on Land Colonization and Rural Credits. *Report of the Commission on Land Colonization and Rural Credits of the State of California.* Sacramento: California State Printing Office, 1916.

California. Department of Agriculture, Division of Land Settlement. *Final Report, June 30, 1931.* Sacramento: California State Printing Office, 1931.

California. Department of Public Works. *Introduction to the First Biennial Report of* the *Division of Land Settlement, Department of Public Works of the State of California, September 1, 1922*. Sacramento: California State Printing Office, 1922.

California. Department of Public Works, Division of Land Settlement. *Farming in Merced County, California: Information for Intending Settlers Regarding the Delhi State Land Settlement in Merced County, California*. Sacramento: California State Printing Office, 1923.

California. Land Settlement Board. *Farm Allotments and Farm Laborers' Allotments in the Delhi State Land Settlement*. Sacramento: California State Printing Office, 1920.

———. *How California Helps Men Own Farms and Rural Homes*. Sacramento: California State Printing Office, 1920.

California Excursion Association. *Southern California, a Semi-Tropical Paradise*. Los Angeles: Times-Mirror, 1887.

"California Takes Lead in Treatment of Disabled Soldiers." *Disabled Veteran Magazine* 1, no. 2 (1921): 21–22.

"California Water and Forest Association." *Forestry and Irrigation* 10, no. 5 (1904): 195; 11, no. 5 (1905): 364–67.

Callinicos, Luli. *Gold and Workers, 1886–1924*. Johannesburg: Ravan, 1985.

Cape of Good Hope. *Douglas Irrigation Works on Vaal River, Griqualand West*. By William Craig. Cape Town: Richards, 1899.

———. *Report on Irrigation Legislation and Enterprise in the American States, Australian Colonies, and Several Countries of Europe as Applicable to Cape Colony*. By William Hammond Hall. Cape Town: Richards, 1898.

———. *Report of the Thebus Irrigation Commission, 1899*. Cape Town: Richards, 1899.

Cape of Good Hope. Department of Public Works. *Report upon Irrigation in South Africa*. By W. Willcocks. Cape Town: Cape Times, 1902.

———. *Report upon the Grobbelaar's River Irrigation Scheme (Oudtshoorn Division)*. Cape Town: Richards, 1896.

Carlson, Martin E. "William E. Smythe: Irrigation Crusader." *Journal of the West* 7, no. 1 (1968): 41–47.

Carlson, Wallin J. "A History of the San Francisco Mining Exchange." Master's thesis, University of California, 1941.

Carruthers, Jane. *The Jameson Raid: A Centennial Retrospective*. Houghton, S.A.: Brenthurst, 1996.

Carter, T. Lane. "The Chinese on the Rand." *Transactions of the American Institute of Mining Engineers* 39 (September 1908): 553–77.

Cartwright, Alan P. *Gold Paved the Way: The Story of the Gold Fields Group of Companies*. London: Macmillan, 1967.

Chaffey Brothers. *The Australian Irrigation Colonies on the River Murray, in Victoria and South Australia*. London: Unwin, 1888–89.

Chan, Sucheng. *This Bittersweet Soil: The Chinese in California Agriculture, 1860–1910*. Berkeley: University of California Press, 1986.

*A Choice Bit of the Land of Sunshine, Fruit, and Flowers. Ontario, Southern California*. Ontario, Calif.: Blackburn, 1896.

Clarke, Frank. *Californian Notes: A Report, Made to the Hon. F. Hagelthorn, on Immigration, Irrigation, Fruit-Growing, and Trade. With an Introduction by Dr. Elwood Mead*. Melbourne: Government Printer, 1915.

Claybourn, John G. *Dredging on the Panama Canal.* Chicago: Kroch, 1931.

Clements, Kendrick A. "Engineers and Conservationists in the Progressive Era." *California History* 58, no. 4 (1979–80): 282–303.

———. *Hoover, Conservation, and Consumerism: Engineering the Good Life.* Lawrence: University Press of Kansas, 2000.

Commager, Henry Steele, ed. *Documents of American History.* 7th ed. New York: Meredith, 1963.

Condé, Jesse C., and Gerald M. Best. *Sugar Trains: Narrow Gauge Rails of Hawaii.* Felton, Calif.: Glenwood, 1973.

Conkin, Paul R. "The Vision of Elwood Mead." *Agricultural History* 34, no. 2 (1960): 88–97.

Conley, Bernice Bedford. *Dreamers and Dwellers: Ontario and Neighbors.* Montclair, Calif.: Stump, 1982.

"Controlling the Colorado River and the Salton Sea." *Scientific American*, December 22, 1906, 467–69.

Cooke, George Paul. *Moolelo O Molokai: A Ranch Story of Molokai.* Honolulu: Honolulu Star-Bulletin, 1949.

Cordray, William Woodrow. "Claus Spreckels of California." Ph.D. diss., University of Southern California, 1955.

Cox, Doak C. "Water Development for Hawaiian Sugar Cane Irrigation." *Hawaiian Planters' Record* 54 (1954): 175–97.

Culcheth, W. W. *Irrigation no. II: Read before the Victorian Institute of Surveyors, 28th April, 1886.* Melbourne: Detmold, 1886.

Curle, J. H. *The Gold Mines of the World.* 2nd ed. London: Waterlow, 1902.

Curti, Merle, and Kendall Birr. *Prelude to Point Four: American Technical Missions Overseas, 1838–1938.* Madison: University of Wisconsin Press, 1954.

Cutliffe, Stephen H., ed. *In Context: History and the History of Technology.* London: Associated University Presses, 1989.

Daniels, Roger. *The Politics of Prejudice: The Anti-Japanese Movement in California and the Struggle for Japanese Exclusion.* Berkeley: University of California Press, 1962.

Davenport, Charles B. *Biographical Memoir of George Davidson, 1825–1911.* Washington, D.C.: National Academy of Sciences, 1937.

Davidson, George. "The Application of Irrigation to California" (1879). In Alonzo Phelps, *Contemporary Biography of California's Representative Men, with Contributions from Distinguished Scholars and Scientists.* Vol. 1. San Francisco: Bancroft, 1881.

Davis, Clark, and David Igler, eds. *The Human Tradition in California.* Wilmington, Del.: Scholarly Resources, 2002.

Davis, Richard. *The English Rothschilds.* Chapel Hill: University of North Carolina Press, 1983.

Deakin, Alfred. *Guide to Irrigation Acts, 1886–1889, with Departmental Regulations, Instructions, Etc.* Melbourne: Government Printer, 1890.

———. *Irrigated India: An Australian View of India and Ceylon, Their Irrigation and Agriculture.* London: Thacker, 1893.

———. *Water Supply and Irrigation: Speech of the Honorable A. Deakin, Chief Secretary of Victoria, in Submitting to the Legislative Assembly a Bill to Make Better Provision for the Supply of Water for Irrigation . . . June 24th, 1886.* Melbourne: Government Printer, 1886.

deBuys, William, and Joan Myers. *Salt Dreams: Land and Water in Low-Down California.*
Albuquerque: University of New Mexico Press, 1999.

"Developing Irrigated Land with Selected Settlers." *Engineering News-Record,*
December 5, 1918.

Deverell, William, and Tom Sitton, eds. *California Progressivism Revisited.* Berkeley:
University of California Press, 1994.

de Waal, Enid. "American Technology in South African Gold Mining before 1899."
*Optima* 33, no. 2 (1985): 81–85.

Dow, John L. *Our Land Acts and a Land Tax; or, The Land Question and How to Deal with
It.* Melbourne: Dunn and Collins, 1877.

Dunlap, Thomas R. *Nature and the English Diaspora: Environment and History in the
United States, Canada, Australia, and New Zealand.* Cambridge: Cambridge University
Press, 1999.

Dupree, A. Hunter. *Science in the Federal Government: A History of Policies and Activities
to 1940.* Cambridge: Harvard University Press, 1957.

Du Puy, William Atherton. *Hawaii and Its Race Problem.* Washington, D.C.: U.S.
Government Printing Office, 1932.

Dyer, Marion Gordon. "The Development of the Beet-Sugar Industry in California."
Master's thesis, University of California, n.d.

East, Lewis R. *Irrigation and Water Supply Development in Victoria.* Melbourne: State
Rivers and Water Supply Commission, 1936.

———. *Victorian Water Law: Riparian Rights.* Melbourne: State Rivers and Water
Supply Commission, n.d.

Easton, Herbert E. *Some Particulars of the Durham, California, Community Land
Settlement.* Sydney: Easton, 1919.

*Eighth Census of the United States, 1860.* Washington, D.C.: U.S. Government Printing
Office, 1864.

Ellsworth, Clayton S. "Theodore Roosevelt's Country Life Commission." *Agricultural
History* 34, no. 4 (1960): 155–72.

Emmet, Boris. *The California and Hawaiian Sugar Refining Corporation.* Stanford, Calif.:
Stanford University Press, 1928.

*Etiwanda: 100 Years.* Etiwanda, Calif.: n.p., 1982.

*Excerpts from the Report of the Royal Commission Appointed by King Edward VIII, to Study
the Problem of Palestine.* New York: National Committee for a Leon Blum Colony in
Palestine, 1937.

"An Extensive Irrigating Project." *Scientific American,* January 27, 1883, 50.

Finch, Charles H. "Workings of a Manager's Mind." *Mining and Scientific Press,*
February 1, 1902.

Fitchett, W. H. "The Australian View of the South African Crisis." *National Review,*
August 1899, 992–99.

Fitzsimmons, W. C. "Colony Building: Colonial Settlements in the Arid West."
*Irrigation Age* 6, no. 2 (1894): 127–28.

Fortier, Samuel. "Field Work in Irrigation: Work of the Office of Experiment Stations
in California in 1905." *Forestry and Irrigation* 11, no. 8 (1905): 371–76.

Friedlander, T. C. "Water and Forest Association: An Organization of Public Spirited

Citizens That Has Done Much for the Higher Development of California's Resources." *Forestry and Irrigation* 11, no. 8 (1905): 365–66.

Friedman, Thomas L. *The World Is Flat: A Brief History of the Twenty-first Century.* New York: Farrar, Straus, and Giroux, 2005.

Galbraith, John S. *Crown and Charter: The Early Years of the British South Africa Company.* Berkeley: University of California Press, 1974.

Gentilcore, R. Louis. "Ontario, California, and the Agricultural Boom of the 1880s." *Agricultural History* 34, no. 2 (1960): 77–87.

Gerber, E., H. G. Prout, and C. C. Schenider. "Memoir of George Shattuck Morison." *Transactions of the American Society of Civil Engineers* 54 (1905): 172–80.

Gilmartin, David. "Scientific Empire and Imperial Science: Colonialism and Irrigation Technology in the Indus Basin." *Journal of Asian Studies* 53, no. 4 (1994): 1127–49.

*The Golden Wheat Lands of Australia.* Sydney: Government Printer, 1910.

"Gold in South Africa." *Mining and Scientific Press,* July 21, 1894.

Gordon, Elizabeth. *What We Saw at Madame World's Fair.* San Francisco: Levinson, 1915.

Griffiths, Tom, and Libby Robin, eds. *Ecology and Empire: Environmental History of Settler Societies.* Seattle: University of Washington Press, 1997.

Grunsky, C. E. "The Lower Colorado River and the Salton Basin." *Transactions of the American Society of Civil Engineers* 59 (1907): 1–62.

———. "The Material Progress of California." *Overland Monthly* 29 (1897): 516–19.

Gullett, H. S. *Irrigation and Intense Culture in Other Countries.* Melbourne: Kemp, 1910.

Gurevich, David. *Fifteen Years of Jewish Immigration into Palestine, 1919–1934.* Jerusalem: Weiss, 1935.

———. *Jewish Agriculture and Agricultural Settlement in Palestine.* Jerusalem: Department of Statistics of the Jewish Agency for Palestine, 1947.

Gutleben, Dan. *Sugar Tramp, 1950, "Hawaiian Issue."* San Francisco: n.p., 1950.

Hall, William Hammond. *Fifth Progress Report of the State Engineer to the Legislature of California, Session of 1887.* Sacramento: California State Printing Office, 1887.

———. *In the Matter of Water Storage and Utilization on the Tuolumne River, California.* San Francisco: Pernau, 1907.

———. *Irrigation Development: History, Customs, Laws, and Administrative Systems Relating to Irrigation, Water-Courses, and Waters in France, Italy, and Spain.* Sacramento: California State Printing Office, 1886.

———. *Irrigation in [Southern] California: The Field, Water-Supply, and Works, Organization and Operation in San Diego, San Bernardino, and Los Angeles Counties. The Second Part of the Report of the State Engineer of California on Irrigation and the Irrigation Question.* Sacramento: California State Printing Office, 1888.

———. *The Irrigation Investigation by the State Engineering Department: Correspondence Concerning the Publication of Its Results.* Sacramento: California State Printing Office, 1882.

———. "Irrigation Principles I: Water Rights." *Irrigation Age* 7, no. 5 (1894): 221–24.

———. "Irrigation Principles II: Association—Irrigation Districts." *Irrigation Age* 7, no. 6 (1894): 252–54.

———. *The Irrigation Question: Memorandum no. 1–2: The Irrigation Question in California: California and Australia.* Sacramento: California State Printing Office, 1886.

———. *The Irrigation Question: Memorandum: Transmitted to the Legislature of California.* Sacramento: California State Printing Office, 1886.

———. *The Irrigation Question in California: Synopsis of a Lecture Delivered on This Subject in the Assembly Chamber.* Sacramento: California State Printing Office, 1878.

———. *Memorandum Concerning the Improvement of the Sacramento River, Addressed to James B. Eads and George H. Mendell.* Sacramento: California State Printing Office, 1880.

———. *Mr. W. Ham. Hall's Statement before the Industrial Commission of Inquiry.* Johannesburg: Johannesburg Times, 1897.

———. *The Office of State Engineer.* Sacramento: Office of State Engineer, 1887.

———. *Physical Data and Statistics of California.* Sacramento: California State Printing Office, 1886.

———. *Proceedings of the State Board of Drainage Commissioners, at a Meeting Held May 28, 1880: Formation of Drainage District no. 1.* Sacramento: California State Printing Office, 1880.

———. *Report of the State Engineer of California: On Irrigation and the Irrigation Question.* 2 vols. Sacramento: California State Printing Office, 1886–88.

———. *Report of the State Engineer to His Excellency R. W. Waterman, Governor of California, for the Year Ending December 31, 1888.* Sacramento: California State Printing Office, 1888.

———. *Report of the State Engineer to the Legislature of the State of California, Session of 1880[-1881].* Sacramento: California State Printing Office, 1880–81.

———. *Second Report of the State Engineer to the Legislature of the State of California, Session of 1881.* Sacramento: California State Printing Office, 1881.

———. *A Statement of the Panama Canal Engineering Conflicts [and] a Review of the Panama Canal Projects: Addressed to the Hon. George C. Perkins.* San Francisco: Philips and Van Orden, 1905.

———. "Statement of William H. Hall, of the U.S. Geological Survey, Supervising Engineer for the Pacific Coast, U.S. Irrigation Survey." In U.S. Senate, *Report of the Special Committee of the United States Senate on the Irrigation and Reclamation of Arid Lands.* Vol. 2, *The Great Basin Region and California,* 208–25. Washington, D.C.: U.S. Government Printing Office, 1890.

———. *A Study of Panama Canal Plans and Arguments: Addressed to the Hon. George C. Perkins.* San Francisco: n.p., 1905.

Hall, William Hammond, Frederick Law Olmsted, and John McLaren. *The Development of Golden Gate Park . . .* San Francisco: Bacon, 1886.

"Hamilton Smith." *Engineering and Mining Journal,* July 14, 1900, 273.

"Hamilton Smith." *Engineering and Mining Journal,* July 28, 1900, 413.

Hammond, John Hays. *The Autobiography of John Hays Hammond.* 2 vols. New York: Farrar and Rinehart, 1935.

———. *The Engineer.* New York: Scribner's, 1921.

———. "The Engineer in Public Life." *Mining and Metallurgy* 10, no. 265 (1929): 115.

———. "The Gold Mines of the Witwatersrand, South Africa." *Engineering Magazine* 14, no. 5 (1898): 733–51.

———. "Gold-Mining in the Transvaal, South Africa." *Transactions of the American Institute of Mining Engineers* 31 (February 1901): 817–54.

————. *The Transvaal Trouble: An Address.* New York: Abbey, 1900.

————. *The Truth about the Jameson Raid.* Boston: Jones, 1918.

Hammond, Natalie Harris. *A Woman's Part in a Revolution.* New York: Longmans, Green, 1897.

Hanson, Warren D. *San Francisco Water and Power.* 3rd ed. San Francisco: City and County of San Francisco, 1994.

Harris, D. G. *Irrigation in India.* London: Oxford University Press, 1923.

Hays, James B. *TVA on the Jordan: Proposals for Irrigation and Hydro-Electric Developments in Palestine.* Washington, D.C.: Public Affairs Press, 1948.

Hays, Samuel P. *Conservation and the Gospel of Efficiency.* Cambridge: Harvard University Press, 1959.

Headrick, Daniel R. *The Tentacles of Progress: Technology Transfer in the Age of Imperialism, 1850–1940.* New York: Oxford University Press, 1988.

————. *The Tools of Empire: Technology and European Imperialism in the Nineteenth Century.* New York: Oxford University Press, 1981.

Herzl, Theodor. *The Jewish State.* 1896; New York: Scopus, 1943.

Hickcox, Robert L., ed. *A History of Etiwanda.* Rancho Cucamonga, Calif.: Community Service Department, 1981.

Hill, Ernestine. *Water into Gold.* 1937; London: Horwitz, 1969.

Hine, Robert V. *California's Utopian Colonies.* New Haven: Yale University Press, 1966.

Hinton, Richard J. *Irrigation in the United States.* 2nd ed. Senate Miscellaneous Document 14, 49th Cong. In U.S. Senate, *Report of the Special Committee of the United States Senate on the Irrigation and Reclamation of Arid Lands,* 265–328. Washington, D.C.: U.S. Government Printing Office, 1890.

Hittell, John S. *The Commerce and Industries of the Pacific Coast of North America.* San Francisco: Bancroft, 1882.

Hobsbawm, Eric. *The Age of Capital, 1848–1875.* New York: Vintage, 1975.

————. *The Age of Empire, 1875–1914.* New York: Vintage, 1987.

Hole, Hugh Marshall. *The Jameson Raid.* London: Allan, 1930.

Holliday, J. S. *The World Rushed In.* New York: Simon and Schuster, 1981.

Horowitz, David. *Jewish Colonisation in Palestine.* Jerusalem: Institute of Economic Research, Jewish Agency for Palestine, 1937.

Huber, Walter L. *An Engineering Century in California.* New York: American Society of Civil Engineers, 1953.

Hughes, Robert. *The Fatal Shore: The Epic of Australia's Founding.* New York: Vintage, 1986.

Hundley, Norris, Jr. *The Great Thirst: Californians and Water, 1770s–1990s.* Berkeley: University of California Press, 1992.

————. "The Politics of Reclamation: California, the Federal Government, and the Origins of the Boulder Canyon Act: A Second Look." *California Historical Quarterly* 52, no. 4 (1973): 292–325.

Igler, David. "Industrial Cowboys: Corporate Ranching in Late Nineteenth-Century California." *Agricultural History* 69, no. 2 (1995): 201–16.

————. *Industrial Cowboys: Miller and Lux and the Transformation of the Far West, 1850–1920.* Berkeley: University of California Press, 2001.

————. "When Is a River Not a River?" *Environmental History* 1, no. 2 (1996): 52–69.

*Irrigation in America: A Series of Articles Written by the Special Commissioner of "The Age."* Melbourne: Kemp, 1910.

*Irrigation in California: The San Joaquin and Tulare Plains.* Sacramento: Record, 1873.

"Irrigation in India." *Forestry and Irrigation* 9, no. 7 (1903): 276–77.

"Irrigation in India." *Scientific American*, December 15, 1883, 279.

"Irrigation in India." *Scientific American*, August 5, 1893, 89–90.

"Irrigation of Australia." *Forestry and Irrigation* 14, no. 4 (1908): 189.

"The Irrigation Works of India." *Scientific American*, August 28, 1875, 128.

Irvine, Leigh H. *The Wedding of the Seas.* Seattle: Crown, 1911.

Isenberg, Andrew C. *Mining California: An Ecological History.* New York: Hill and Wang, 2005.

Islam, M. Mufakharul. *Irrigation, Agriculture, and the Raj: Punjab, 1887–1947.* Manohar, India: South Asia Institute, 1997.

Jackson, Donald C. *Building the Ultimate Dam: John S. Eastwood and Control of Water in the West.* Lawrence: University Press of Kansas, 1995.

James, George Wharton. "The Overflow of the Colorado and the Salton Sea: II." *Scientific American*, April 21, 1906, 328–29.

Jenkins, Julian. "The Boer: By One Who Knows Him." *Westminster Review*, June 1900, 646–51.

Jha, U. M. *Irrigation and Agricultural Development.* New Delhi: Deep and Deep, 1984.

Johnson, Alvin. "Land for the Returned Soldier." *New Republic*, September 21, 1919, 218–20.

Johnson, J. W. "Early Engineering in California." *California Historical Society Quarterly* 29, no. 3 (1950): 193–209.

Johnson, Paul. *Consolidated Gold Fields: A Centenary Portrait.* New York: St. Martin's, 1987.

Joint Palestine Survey Commission, Frank Adams, Sir John Campbell, Cyril Q. Henriques, Jacob Goodale Lipman, Elwood Mead, Milton Joseph Rosenau, Sir Edward J. Russell, Knowles Augustus Ryerson, Arthur Thomas Strahorn, Charles Francis Wilinsky, and Leo Wolman. *Reports of the Experts Submitted to the Joint Palestine Survey Commission.* Boston: Daniels, 1928.

Jones, Robert E. "Ready-Made Farms." *Country Gentleman*, July 6, 1918.

*Journal of the Assembly during the Twenty-second Session of the Legislature of the State of California, 1877–78.* Sacramento: California State Printing Office, 1878.

Kahrl, William L. *The California Water Atlas.* Sacramento: California Department of Water Resources, 1979.

———. *Water and Power: The Conflict over Los Angeles' Water Supply in the Owens Valley.* Berkeley: University of California Press, 1982.

Kamen, Charles S. *Little Common Ground: Arab Agriculture and Jewish Settlement in Palestine, 1920–1948.* Pittsburgh: University of Pittsburgh Press, 1991.

Karlinsky, Nahum. "California Dreaming: Adapting the 'California Model' to the Jewish Citrus Industry in Palestine, 1917–1939." *Israel Studies* 5, no. 1 (2000): 24–40.

Katz, Elaine. "Revisiting the Origins of the Industrial Colour Bar in the Witwatersrand Gold Mining Industry, 1891–1899." *Journal of Southern African Studies* 25, no. 1 (1999): 71–95.

Kauanui, J. Kehaulani. "Rehabilitating the Native: Hawaiian Blood Quantum and the

Politics of Race, Citizenship, and Entitlement." Ph.D. diss., University of California, 2000.

Keesing, Felix M. *Hawaiian Homesteading on Molokai.* Honolulu: University of Hawaii, 1936.

Kelley, Robert L. *Battling the Inland Sea: Floods, Public Policy, and the Sacramento Valley.* Berkeley: University of California Press, 1989.

———. *Gold versus Grain: The Hydraulic Mining Controversy in California's Sacramento Valley.* Glendale, Calif.: Clark, 1959.

Kenyon, A. S. "Stuart Murray and Irrigation in Victoria." *Victorian Historical Magazine,* July 25, 1921, 112–22.

Kershner, Frederick D. "George Chaffey and the Irrigation Frontier." *Agricultural History* 27, no. 4 (1953): 115–22.

Kieffer, Stephen E., C. E. Grunsky, and J. B. Lippincott. "Memoir of James Dix Schuyler." *Transactions of the American Society of Civil Engineers* 76 (1913): 2243–45.

Kinsley, Guy R. "Panama-Pacific Exposition Triumphs over the Effects of European War." *Out West* 8, no. 5 (1914): 225.

Kluger, James R. *Turning on Water with a Shovel: The Career of Elwood Mead.* Albuquerque: University of New Mexico Press, 1992.

Korr, Charles P. "William Hammond Hall: The Failure of Attempts at State Water Planning in California, 1878–1888." *Southern California Quarterly* 45, no. 4 (1963): 305–22.

Kubicek, Robert V. *Economic Imperialism in Theory and Practice: The Case of South African Gold Mining Finance, 1886–1914.* Durham: Duke University Press for the Duke University Center for Commonwealth and Comparative Studies, 1979.

Kumar, Dharma, ed. *The Cambridge Economic History of India.* Vol. 2. Cambridge: Cambridge University Press, 1984.

"Labor Conditions in Johannesburg." *Engineering and Mining Journal,* June 12, 1902, 951.

"Labor Conditions in the Transvaal Mines." *Engineering and Mining Journal,* July 12, 1902, 1042.

LaFeber, Walter. *The New Empire: An Interpretation of American Expansion, 1860–1898.* Ithaca: Cornell University Press, 1963.

Lavender, David. *California: Land of New Beginnings.* Lincoln: University of Nebraska Press, 1972.

———. *Nothing Seemed Impossible: William C. Ralston and Early San Francisco.* Palo Alto, Calif.: American West, 1975.

Layton, Edwin T., Jr. *The Revolt of the Engineers: Social Responsibility and the American Engineering Profession.* Baltimore: Johns Hopkins University Press, 1986.

Lee, Lawrence B. "William Ellsworth Smythe and the Irrigation Movement: A Reconsideration." *Pacific Historical Review* 41, no. 3 (1972): 289–311.

Leroy-Beaulieu, Pierre. "The Labor Question in the Transvaal Mines." *Engineering and Mining Journal,* March 14, 1896.

Lewis, Oscar. *George Davidson: Pioneer West Coast Scientist.* Berkeley: University of California Press, 1954.

Liebman, Ellen. *California Farmland: A History of Large Agricultural Landholdings.* Totowa, N.J.: Rowman and Allanheld, 1983.

*Life and a Living.* Ontario, Calif.: Ontario Chamber of Commerce, n.d.

Lippincott, J. B. "General Outlook for Reclamation Work in California." *Forestry and Irrigation* 11, no. 8 (1905): 349–53.

———. "The Reclamation Service in California." *Forestry and Irrigation* 10, no. 4 (1904): 162–69.

Lockhart, J. G., and C. M. Woodhouse. *Cecil Rhodes: The Colossus of Southern Africa.* New York: Macmillan, 1963.

Lowdermilk, Walter Clay. *Palestine: Land of Promise.* New York: Harper, 1944.

Maass, Arthur, and Raymond L. Anderson. *. . . and the Desert Shall Rejoice: Conflict, Growth, and Justice in Arid Environments.* Cambridge: MIT Press, 1978.

MacKenzie, Melody, ed. *Native Hawaiian Rights Handbook.* Honolulu: Native Hawaiian Legal Corporation and the Office of Hawaiian Affairs, 1993.

Macnab, Roy. *Gold Their Touchstone: Gold Fields of South Africa 1887–1987, a Centenary Story.* Johannesburg: Ball, 1987.

Macomber, Ben. *The Jewel City.* San Francisco: Williams, 1915.

Malone, Dumas, ed. *Dictionary of American Biography.* Vols. 9, 17, 20. New York: Scribner's, 1935.

Marutani, Herbert K. "Labor-Management Relations in Agriculture: A Study of the Hawaiian Sugar Industry." Ph.D. diss., University of Hawaii, 1970.

Marx, Karl, and Friedrich Engels. *Manifesto of the Communist Party.* 1848; Chicago: Kerr, 1906.

Maxwell, George H. "Annex Arid America." *Irrigation Age,* November 1898, 51–55.

———. "Reclamation of Arid America." *Irrigation Age,* September 1899, 407–9.

McColl, James H. *Agriculture and Irrigation: Information Collected by Hon. J. H. McColl, M.P., during His Recent Trip through America.* Bendigo, Australia: Cambridge, 1906.

McCoy, C. G. *Victorian Irrigation and Drainage Practice: Paper 1: Historical Development of Irrigation in Victoria.* Melbourne: State Rivers and Water Supply Commission, [198?].

McNeill, John R. *Something New under the Sun: An Environmental History of the Twentieth-Century World.* New York: Norton, 2000.

McPherson, Alex. "The Duty of Water." *Forestry and Irrigation* 12, no. 9 (1906): 417.

McWilliams, Carey. *California: The Great Exception.* New York: Current, 1949.

———. *Factories in the Fields.* Boston: Little, Brown, 1939.

Mead, Elwood. *Agricultural Development in Palestine: Report to the Zionist Executive.* 2nd ed. London: Zionist Executive, 1924.

———. *Colonization and Rural Development in California.* Berkeley: University of California, College of Agriculture, Agricultural Experiment Station, 1920.

———. *Comparison of Land Settlement Methods in California with Those in Other Countries.* Berkeley: University of California, 1916.

———. "Farming in Palestine: Preparing Young Women for It." *New Palestine,* May 23, 1924, 416–17.

———. "Government Aid and Direction in Land Settlement." *American Economic Review* 8, supplement (1918): 89–91.

———. "The Growth of Irrigation in America: The Accomplishments of Scientific Farming in the West." *Irrigation Age* 14, no. 11 (1900): 376–85.

———. *Helping Men Own Farms: A Practical Discussion of Government Aid in Land Settlement.* New York: Macmillan, 1920.

———. "Irrigation in California." *Irrigation Age* 14, no. 4 (1900): 121–24.

———. "Irrigation in Victoria." *Engineering Record*, August 14, 1909, 175–76.

———. "Irrigation in Victoria." In *Handbook to Victoria*, ed. A. M. Laughton and T. S. Hall, 255–68. Melbourne: British Association for the Advancement of Science, 1914.

———. *Irrigation Institutions: A Discussion of the Economic and Legal Questions Created by the Growth of Irrigated Agriculture in the West*. New York: Macmillan, 1903.

———. "The Irrigation Investigations in California of the Office of Experiment Stations." *Forestry and Irrigation* 11, no. 8 (1905): 367–69.

———. *Land Settlement and Rural Credits: Statement of the Need for an Investigation*. Berkeley: University of California, 1916.

———. *Memorandum by Dr. Elwood Mead Respecting Allocation of Murray Waters, and Amendments to Laws Governing Water Rights*. Sydney: Kent, 1923.

———. "The New Forty-niners." *Survey Graphic* 1, no. 4 (1922): 651–58.

———. "The New Palestine." *American Review of Reviews* 70 (December 1924): 623–29.

———. *Report of Investigations of Land Settlement and Irrigation Development in America*. Melbourne: Mullett, 1914.

———. *Report of Irrigation Investigations in California, under the Direction of Elwood Mead, Assisted by William E. Smythe, Marsden Manson, J. M. Wilson, Charles D. Marx, Frank Soule, C. E. Grunsky, Edward M. Boggs . . .* Washington, D.C.: U.S. Government Printing Office, 1901.

———. *Report of the Division of Land Settlement, a Subdivision of the Department of Public Works of the State of California . . . September 1, 1922*. Sacramento: California State Printing Office, 1922.

———. *Report on the Murrumbidgee Irrigation Scheme*. Sydney: Government Printer, 1923.

———. *The Rural Credit System Needed in Western Development*. Berkeley: University of California, 1916.

———. *State Aid in Land Settlement: An Address by Elwood Mead . . .* Sacramento: California State Printing Office, 1915.

———. "The Tenant Farmer and Land Monopoly." In *Proceedings, National Conference of Social Work*, 1–5. Chicago: Roberts and Hall, 1918.

———. "What Australia Can Teach America." *Independent*, August 17, 1911, 367–70.

Mead, Elwood, Frank Adams, World Zionist Organization, and Joint Palestine Survey Commission. *Agricultural Colonization in Palestine*. N.p., 1927.

Meinig, Donald W. "The Growth of Agricultural Regions in the Far West: 1850–1910." *Journal of Geography* 54, no. 5 (1955): 221–32.

Mildura Royal Commission. *Mildura Irrigation Settlement: Complaints by Resident Land-Owners: Report to the Hon. J. H. McColl, Minister of Water Supply, by Stuart Murray, Chief Engineer*. Melbourne: Government Printer, 1887–96.

Miller, M. Catherine. *Flooding the Courtrooms: Law and Water in the Far West*. Lincoln: University of Nebraska Press, 1993.

"The Mines of West Australia." *Mining and Scientific Press*, January 19, 1895.

Mintner, William. *King Solomon's Mines Revisited: Western Interests and the Burdened History of Southern Africa*. New York: Basic Books, 1986.

Moeller, E. O. *The Ontario-Cucamonga-Etiwanda Colonies: The Banner Fruit Districts of Southern California*. San Francisco: Stanley-Taylor, 1902.

Moore, Charles C. "San Francisco and the Exposition." *Sunset*, 28, no. 2 (1912): 196–98.

———. "'San Francisco Knows How!'" *Sunset* 28, no. 1 (1912): 3–15.

Morison, George S. "Address at the Annual Convention at the Hotel Pemberton, Hull, Mass., June 19, 1895." *Transactions of the American Society of Civil Engineers* 33, no. 751 (1895): 467–84.

Morris, Benny. *Righteous Victims: A History of the Zionist-Arab Conflict, 1881–2001*. New York: Vintage, 2001.

Mumford, Lewis. *Technics and Civilization*. New York: Harcourt, Brace, 1934.

Murray, A. S. *Twelve Hundred Miles on the River Murray*. Melbourne: Robertson, 1898.

Murray, John. *The Victorian Settlers' Guide: Handbook of the Land Acts (Includes "Closer Settlement Act 1904"), Revised and Brought Up to Date*. Melbourne: Government Printer, 1905.

Murray, Stuart. *Irrigation in Victoria: Its Position and Prospects*. Melbourne: Government Printer, 1892.

Myers, Francis. *Irrigation; or, The New Australia*. Melbourne: Chaffey, 1891.

Nash, George H. *The Life of Herbert Hoover: The Engineer, 1874–1914*. New York: Norton, 1983.

Nash, Gerald D. *State Government and Economic Development: A History of Administrative Policies in California, 1849–1933*. Berkeley: Institute of Governmental Studies, University of California, 1964.

"Native Labor in South African Gold Mines." *Mining and Scientific Press*, January 18, 1896.

*A New Eldorado: The Goulburn Valley: A Short History of Its Settlement, Present Development, and Future Prospects*. Melbourne: Mooroopna Irrigable Lands Investment, 1889.

Newell, Frederick H. "Awakening of the Engineer." *Engineering News*, September 16, 1915, 568–70.

———. "The Engineer in the Public Service." *Engineering News*, July 25, 1912, 153–55.

———. "Future Progress in Irrigation." *Engineering News*, December 5, 1912, 1036–41.

———. *Hawaii: Its Natural Resources and Opportunities for Home-Making*. Washington, D.C.: U.S. Government Printing Office, 1909.

———. "Work of the Reclamation Service in California." *Forestry and Irrigation* 11, no. 8 (1905): 346–47.

"New Gold Fields in South Africa." *Mining and Scientific Press*, December 15, 1894.

Newlands, Francis G. "National Irrigation as a Social Problem." *Pacific Monthly* 16, no. 3 (1906): 295–97.

———. "National Irrigation Works." *Forestry and Irrigation* 8, no. 2 (1902): 63–66.

Nordhoff, Charles. *Northern California, Oregon, and the Sandwich Islands*. 1874; Berkeley: Ten Speed, 1974.

Norman, Theodore. *An Outstretched Arm: A History of the Jewish Colonization Association*. London: Routledge and Kegan Paul, 1985.

Norton, C. E. "Irrigation in India." *North American Review* 77 (1855).

Norton, Charles P. "Where Will Our Soldiers Go after the War?" *New West Magazine* 9, no. 10 (1918): 17–19.

Nye, David E. "Electrifying the West, 1880–1940." In *The American West as Seen by Europeans and Americans*, ed. R. Kroes, 183–202. Amsterdam: Free University Press, 1989.

Oettinger, Jacob. *Jewish Colonization in Palestine: Methods, Plans, and Capital.* The Hague: Head Office of the Jewish National Fund, [191?].

*Ontario: Located in San Bernardino County, California, on the Southern Pacific Railroad . . .* Riverside, Calif.: Press and Horticulturist, 1883.

*Ontario Colony, California, Comprising Ontario, Upland, San Antonio Heights.* Ontario, Calif.: Chamber of Commerce, 1904.

Orsi, Richard J. "The Octopus Reconsidered: The Southern Pacific and Agricultural Modernization in California, 1865–1915." *California Historical Quarterly* 54 (Fall 1975): 197–220.

Osborne, Sherrill B. *Ontario Colony, California.* Ontario, Calif.: Ontario Chamber of Commerce, [190?].

O'Shaughnessy, Michael M. "Irrigation Works in the Hawaiian Islands." *Transactions of the American Society of Civil Engineers* 54, pt. C (1905): 129–36.

"The Panama Canal." *Transactions of the American Institute of Mining Engineers* 49 (January 1911): 1–84.

"The Panama-Pacific Exposition Comes to an End." *Engineering News,* December 9, 1915, 1138.

Panama-Pacific International Exposition Company. Committee on Exploration and Publicity. *San Francisco, the Exposition City: The West Invites the World.* San Francisco: Merchants Exchange, 1915.

Panama Pacific International Exposition Company. Mining Week Committee. *Souvenir Program, Mining Week.* San Francisco: Panama Pacific International Exposition, 1915.

"The Passing of the Pioneer." *Mining and Scientific Press,* April 3, 1915, 507.

Paul, Rodman W. *Mining Frontiers of the Far West, 1848–1880.* New York: Holt, Rinehart, and Winston, 1963.

Pearlman, Maurice. *Collective Adventure: An Informal Account of the Communal Settlements of Palestine.* London: Heinemann, 1937.

Penslar, Derek. "Engineering Utopia: The World Zionist Organization and the Settlement of Palestine, 1897–1914." Ph.D. diss., University of California, 1987.

Perkins, Horace. *Melbourne Illustrated and Victoria Described: The Visitors Handbook of Facts and Figures.* Melbourne: for the author, [189?].

Perrucci, Robert, and Joel E. Gerstl, eds. *The Engineers and the Social System.* New York: Wiley, 1969.

Phelps, Alonzo. *Contemporary Biography of California's Representative Men.* San Francisco: Bancroft, 1882.

Pickstone, H. E. V. *Report of Mr. H. E. V. Pickstone in Regard to the Forming of the Affairs of the Rhodes Fruit Farms, Etc. into a Limited Liability Company under the Style of "Rhodes Limited."* Cape Town: n.p., 1901.

Pinchot, Gifford. "Relation of Forests to Irrigation." *Forestry and Irrigation* 10, no. 12 (1904): 551–52.

Pisani, Donald J. *From the Family Farm to Agribusiness: The Irrigation Crusade in California and the West, 1850–1931.* Berkeley: University of California Press, 1984.

———. *To Reclaim a Divided West: Water, Law, and Public Policy, 1848–1902.* Albuquerque: University of New Mexico Press, 1992.

———. *Water, Land, and Law in the West: The Limits of Public Policy, 1850–1920.* Lawrence: University Press of Kansas, 1996.

———. "Water Law Reform in California, 1900–1913." *Agricultural History* 54, no. 2 (1980): 295–317.

Polk, Willis. "The Panama-Pacific Plan." *Sunset* 28, no. 4 (1912): 487–92.

Pomeroy, Earl. *The Pacific Slope: A History of California, Oregon, Washington, Idaho, Utah, and Nevada.* Lincoln: University of Nebraska Press, 1965.

Porter, Bernard. *The Lion's Share: A Short History of British Imperialism, 1850–1995.* 3rd ed. London: Longman, 1996.

Powell, John Wesley. "Institutions for the Arid Lands." *Century*, May 1890, 111–16.

Powell, Joseph M. "Elwood Mead and California's State Colonies: An Episode in Australasian-American Contacts, 1915–31." *Journal of the Royal Australian Historical Society* 67, no. 4 (1982): 328–53.

———. *Environmental Management in Australia, 1788–1914.* Melbourne: Oxford University Press, 1976.

———. *An Historical Geography of Modern Australia: The Restive Fringe.* Cambridge: Cambridge University Press, 1988.

———. *Watering the Garden State: Water, Land and Community in Victoria, 1834–1988.* Sydney: Allen and Unwin, 1989.

Probert, Frank H. "Mining Alumni, an Appraisal and Appreciation." *California Monthly*, April 1925, 447–51.

Prout, H. G. "The Economic Conquest of Africa." *Engineering Magazine* 18, no. 5 (1900): 657–80.

Pursell, Carroll W. *The Machine in America: A Social History of Technology.* Baltimore: Johns Hopkins University Press, 1995.

———. "The Technical Society of the Pacific Coast, 1884–1914." *Technology and Culture* 17 (October 1976): 707–17.

Raymond, R. W. "Biographical Notice of Louis Janin." *Transactions of the American Institute of Mining Engineers* 9 (July 1914): xxxvii, 1406–10.

Reay, W. T. *Irrigation and Intense Culture in Other Countries.* Melbourne: Kemp, n.d.

*Rehabilitation in Hawaii.* Honolulu: Hawaiian Homes Commission, 1922.

Reid, Bill G. "Agrarian Opposition to Franklin K. Lane's Proposal for Soldier Settlement, 1918–1921." *Agricultural History* 41, no. 2 (1967): 167–79.

*Report of the Board of Commissioners on the Irrigation of the San Joaquin, Tulare, and Sacramento Valleys of the State of California.* 43rd Cong., 1st sess., House Ex. Doc. 290. Washington, D.C.: U.S. Government Printing Office, 1874.

*Report of the Commission on Country Life, with an Introduction by Theodore Roosevelt.* Chapel Hill: University of North Carolina Press, 1944.

*Report of the Hawaiian Homes Commission to the Legislature of Hawaii.* Honolulu: Hawaiian Homes Commission, 1929.

*Reports of the Committee on Beet Sugar Culture in California. Submitted to a Special Meeting of the Chamber of Commerce of San Francisco, Thursday, February 2d, 1888.* San Francisco: Commercial, 1888.

Revusky, Abraham. *Jews in Palestine.* New York: Vanguard, 1945.

Rickard, T. A. *A History of American Mining.* New York: McGraw-Hill, 1932.

———. *Interviews with Mining Engineers.* San Francisco: Mining and Scientific, 1922.

Rickard, T. A., H. C. Hoover, W. B. Ingalls, R. G. Brown, and Other Specialists. *The Economics of Mining.* New York: Engineering and Mining Journal, 1905.

Robbins, William G. *Colony and Empire: The Capitalist Transformation of the American West*. Lawrence: University Press of Kansas, 1994.

Rodgers, Daniel T. *Atlantic Crossings: Social Politics in a Progressive Age*. Cambridge: Belknap Press of Harvard University Press, 1998.

Rook, Robert E. "An American in Palestine: Elwood Mead and Zionist Water Resource Planning, 1923–1936." *Arab Studies Quarterly* 22, no. 1 (2000): 71–90.

Ruppin, Arthur. *The Agricultural Colonisation of the Zionist Organisation in Palestine*. Trans. R. J. Feiwel. London: Hopkinson, 1926.

Rutherford, J. "Interplay of American and Australian Ideas for Development of Water Projects in Northern Victoria." In *The Making of Rural Australia: Environment, Society, and Economy*, ed. Joseph M. Powell, 116–34. Melbourne: Sorrett, 1974.

Ryerson, Knowles A. *The World Is My Campus*. Davis: Regents of the University of California, 1977.

Sabin, Paul. "Home and Abroad: The Two 'Wests' of Twentieth-Century United States History." *Pacific Historical Review* 66, no. 3 (1997): 305–35.

Sackett, Robert L. *The Engineer: His Work and His Education*. New York: Ginn, 1928.

"The Sale of Anaconda Stock." *Engineering and Mining Journal*, June 13, 1896, 561; September 28, 1895, 294; October 26, 1895, 389; June 13, 1896, 561.

San Joaquin and King's River Canal and Irrigation Company. *Agricultural Lands and Waters in the San Joaquin and Tulare Valleys*. San Francisco: Bancroft, 1873.

———. *The Great San Joaquin and Sacramento Valleys of California: A Great Irrigation Scheme*. N.p., 1873.

———. *Report of the San Joaquin and King's River Canal and Irrigation Company, for 1873. Property, Rights, Progress, Etc. by Order of Stockholders*. San Francisco: Woman's Publishing, 1874.

Saxton, Alexander. *The Indispensable Enemy: A Study of the Anti-Chinese Movement in California*. Berkeley: University of California Press, 1967.

Schlereth, Thomas J. *Victorian America: Transformations in Everyday Life, 1876–1915*. New York: HarperCollins, 1991.

Schorske, Carl E. *Fin-de-Siècle Vienna: Politics and Culture*. 1961; New York: Vintage, 1981.

Schuyler, James D. *Culture of Sugar Cane: Report on Water Supply for Irrigation on the Honouliuli and Kahuku Ranchos, Island of Oahu, Hawaiian Islands*. Oakland, Calif.: Jordan and Arnold, 1889.

———. *Report on the Water Supply, the Projected Storage Works, the Proposed Electric Power Development, and the Prospective Business of the Wahiawa Water Co. Ltd.* Los Angeles: n.p., 1903.

Scott-Moncrieff, Colin. "Irrigation in Egypt." *Nineteenth Century* 17, no. 96 (1885): 342–44.

———. *Irrigation in Southern Europe: Being the Report of a Tour of Inspection of the Irrigation Works of France, Spain, and Italy, Undertaken in 1867–68 for the Government of India*. London: Spon, 1868.

Shariff, Ismail. *The Development of Indian and American Agriculture: A Comparative Study*. Green Bay: University of Wisconsin, Green Bay, n.d.

Shaw, Albert. "California's Farm Colonies." *American Review of Reviews* 64 (1921).

Shaw, George W. *The California Sugar Industry*. Part 1, *Historical and General*. Berkeley: Agricultural Experiment Station, University of California, 1903.

Shepstone, H. J. "The Irrigation Works of India." *Scientific American Supplement*, September 11, 1915, 154–65.

Shillington, Kevin. *A History of Southern Africa*. Essex, England: Longman, 1987.

Shinn, Charles Howard. "California Forests and Irrigation." *Garden and Forest*, September 1890, 426–27.

"Shipping Chinamen to South Africa." *Engineering and Mining Journal*, August 18, 1904, 562.

Skiff, Frederick J. V. "The Neighborhood of the World." *Sunset* 28, no. 5 (1912): 619–25.

Slocum, William F. "The Universities' Interest in Irrigation Problems." *Forestry and Irrigation* 8, no. 11 (1902): 474–75.

Smith, Iain R. *The Origins of the South African War, 1899–1902*. London: Longman, 1996.

Smith, Michael L. *Pacific Visions: California Scientists and the Environment, 1850–1915*. New Haven: Yale University Press, 1987.

Smythe, William E. "The Colony Builders." *Irrigation Age* 14, no. 1 (1899): 19–23.

———. "The Colony-Builders." *Irrigation Age* 14, no. 2 (1899): 60–66.

———. *The Conquest of Arid America*. New York: Harper, 1900.

———. "The Ennoblement of California." *Out West* 24, no. 6 (1906): 539–41.

———. "The Ethics of Irrigation." *Out West* 18, no. 2 (1903): 233–43.

———. "The Homemaker or the Speculator?" *Forestry and Irrigation* 9, no. 9 (1903): 442–46.

———. "Our Great Pacific Commonwealth: A Study of Ultimate California." *Century Magazine*, December 1896, 300–307.

———. "A Program for California." *Land of Sunshine*, December 1901, 487–98.

———. "Real Utopias in the Arid West." *Atlantic Monthly*, May 1897, 599–610.

———. "The Republic of Irrigation." *Irrigation Age* 6, no. 5 (1894): 189–95.

*Something about Sugar*. San Francisco: California and Hawaiian Sugar Refining, 1925.

Spence, Clark C. *British Investments and the American Mining Frontier, 1860–1901*. Ithaca: Cornell University Press, 1958.

———. *Mining Engineers and the American West: The Lace-Boot Brigade, 1849–1933*. New Haven: Yale University Press, 1970.

Spreckels, J. D., and Brothers. *Ports of San Francisco, San Diego, Puget Sound, Portland, and Honolulu*. San Francisco: Brown, 1889.

Springer, Thomas Grant, and Fleta Campbell Springer. "The Water-Way of Wonder." *Sunset* 28, no. 5 (1912): 544–50.

Starr, Kevin. *Americans and the California Dream, 1850–1915*. New York: Oxford University Press, 1973.

———. *Inventing the Dream: California through the Progressive Era*. New York: Oxford University Press, 1985.

———. *Material Dreams: Southern California through the 1920s*. New York: Oxford University Press, 1990.

Stevens, Sylvester K. *American Expansion in Hawaii, 1842–1898*. New York: Russell and Russell, 1968.

*A Souvenir from Etiwanda: The Prize Winning Citrus Colony.* Etiwanda, Calif.: Etiwanda Board of Trade, [191?].

*Story of Sugar in Hawaii.* Honolulu: Hawaiian Sugar Planters' Association, 1926.

Teele, R. P. "The Organization of Irrigation Companies." *Journal of Political Economy* 12, no. 2 (1904): 161–78.

———. *The State Engineer and His Relation to Irrigation.* Washington, D.C.: U.S. Government Printing Office, 1906.

Teisch, Jessica B. "'Home Is Not So Far Away': Californian Engineers in South Africa, 1868–1915." *Australian Economic History Review* 45, no. 2 (2005): 139–60.

———. "Sweetening the Urban Marketplace: California's Hawaiian Outpost." In *Cities and Nature in the American West*, ed. Char Miller. Reno: University of Nevada Press, 2010.

———. "William Hammond Hall, City Water, and Progressive-Era Reform in San Francisco." In *The Human Tradition in California*, edited by Clark Davis and David Igler, 81–98. Wilmington, Del.: Scholarly Resources, 2002.

Thomson, Lloyd, and Sydney Wells, eds. *Renmark Newspaper Slips, 1886–1889.* Mildura: Mildura Legacy Club, 1989.

Thurston, Lorrin A. *The Labor Situation in Hawaii: A Paper Read by Lorrin A. Thurston, before the Social Science Club of Honolulu.* Honolulu: Hawaiian Gazette, 1906.

Todd, Frank Morton. *The Story of the Exposition.* 5 vols. New York: Knickerbocker, 1921.

*Transactions of the State Agricultural Society, 1866–1867.* Sacramento: Government Printing Office, 1868.

*Transactions of the State Agricultural Society, 1870–1871.* Sacramento: Government Printing Office, 1872.

"The Transvaal Crisis." *Engineering and Mining Journal*, January 18, 1896, 59–60.

"The Transvaal Troubles." *Engineering and Mining Journal*, January 25, 1896, 256.

Trask, F. E. "The Irrigation System of Ontario: Its Development and Cost." *Transactions of the American Society of Civil Engineers* 55, no. 175 (1905): 173–82.

Troen, S. Ilan. "Frontier Myths and Their Applications in America and Israel: A Transnational Perspective." *Israel Studies* 5, no. 1 (2000): 301–29.

Tucker, Robert C., ed. *The Marx-Engels Reader.* New York: Norton, 1972.

Turai, Abraham. *The Emek Jezreel and the Beisan Valley.* Tel Aviv: Lidov, 1945.

Turner, Frederick Jackson. "The Significance of the Frontier in American History (1893)." In *The Frontier in American History*, 1–38. New York: Holt, Rinehart, and Winston, 1962.

Turrell, Robert Vicat. "Rhodes, De Beers, and Monopoly." *Journal of Imperial and Commonwealth History* 10, no. 3 (1982): 311–43.

Turrell, Robert Vicat, and Jean-Jacques Van Helten. "The Rothschilds, the Exploration Company, and Mining Finance." *Business History* 28, no. 2 (1986): 181–205.

Twain, Mark. *The Innocents Abroad; or, The New Pilgrims' Progress.* 1869; New York: New American Library, 1980.

———. *Letters from Honolulu Written for the Sacramento Union.* Intro. John W. Vandercook. Honolulu: Nickerson, 1939.

———. *More Tramps Abroad.* London: Chatto and Windus, 1897.

Tyrrell, Ian. "American Exceptionalism in an Age of International History." *American Historical Review* 96, no. 4 (1991): 1031–56.

——. *True Gardens of the Gods: Californian-Australian Environmental Reform, 1860–1930.* Berkeley: University of California Press, 1999.

U.S. Board of Commissioners on the Irrigation of the San Joaquin, Tulare, and Sacramento Valleys of the State of California. *Engineers and Irrigation: Report of the Board of Commissioners on the Irrigation of the San Joaquin, Tulare, and Sacramento Valleys of the State of California, 1873.* 1874; Fort Belvoir, Va.: Office of History, U.S. Army Corps of Engineers, 1990.

U.S. Bureau of Mines. *Exhibits of U.S. Bureau of Mines in Mines and Metallurgy Palace.* San Francisco: Panama-Pacific International Exposition, 1915.

U.S. Coast and Geodetic Survey. *Irrigation and Reclamation of Land for Agricultural Purposes, as Now Practiced in India, Egypt, Italy, etc.* By George Davidson. 44th Cong., 1st sess., Senate Ex. Doc. 94. Washington, D.C.: U.S. Government Printing Office, 1875.

U.S. Department of Agriculture. *Annual Report of the Department of Agriculture for the Fiscal Year Ended June 30, 1900.* Washington, D.C.: U.S. Government Printing Office, 1900.

U.S. Department of Agriculture. Office of Experiment Stations. *Irrigation in Hawaii.* By Walter Maxwell. Washington, D.C.: U.S. Government Printing Office, 1900.

U.S. Department of the Interior. *Summary of Soldier Settlements in English-Speaking Countries.* By Elwood Mead. Washington, D.C.: U.S. Government Printing Office, 1919.

U.S. Geological Survey. *Irrigation Bulletin: Water-Duty Memoranda and Conversion Tables.* By William Hammond Hall. Washington, D.C.: U.S. Government Printing Office, 1890.

"The United States Geological Survey at the Panama Exposition." *Science,* March 12, 1915, 383–84.

U.S. Senate. *Report of the Special Committee of the United States Senate on the Irrigation and Reclamation of Arid Lands.* Part 2, vol. 4, *Reports on Irrigation from Consuls of the United States.* Washington, D.C.: U.S. Government Printing Office, 1890.

Ulitzur, A. *Two Decades of Keren Hayesod.* Jerusalem: Keren Hayesod, 1940.

Ussishkin, Anne. "The Jewish Colonisation Association and a Rothschild in Palestine." *Middle Eastern Studies* 9, no. 3 (1973): 347–57.

Vamplew, Wray, ed. *Australians: Historical Statistics.* New South Wales: Fairfax, Syme, and Weldon, 1987.

Vandercook, John W. *King Cane: The Story of Sugar in Hawaii.* New York: Harper, 1939.

Vause, Marylyn. "The Hawaiian Homes Commission Act, 1920: History and Analysis." Master's thesis, University of Hawaii, 1962.

Veblen, Thorstein. *The Engineers and the Price System.* New York: Huebsch, 1921.

Victoria. *Murray River Waters and Mallee Frontage Settlement. Parliamentary Trip of Inspection by Steam-Boat from Echuca to Mildura and the Darling Junction, October 17th to 22nd, 1907.* Melbourne: Government Printer, 1907.

Victoria. Mildura Settlement. *Report of the Mildura Royal Commission.* Melbourne: Brain, 1896.

Victoria. Rivers and Water Supply Commission. *Advice to Settlers in the Irrigation Districts.* Melbourne: Government Printer, 1913.

———. *Irrigation and Water Supply Development in Victoria.* Comp. H. L. Boorman. Melbourne: Gourley, 1947.

———. *Irrigation Trusts Blue Books, 1884–1894.* 2 vols. Melbourne: Rivers and Water Supply Commission, 1884–94.

Victoria. Royal Commission on Water Supply. *First Progress Report: Irrigation in Western America, so Far as It Has Relation to the Circumstances of Victoria.* By Alfred Deakin. Melbourne: Government Printer, 1885.

———. *Further Progress Report: Presented to Both Houses of Parliament by His Excellency's Command.* By Alfred Deakin. Melbourne: Government Printer, 1885.

Victorian Water Supply Department. *Proceedings of the First Conference of Irrigationists, Held in Melbourne, on 25th, 26th, and 28th March, 1890.* Melbourne: Government Printer, 1890.

Vincent, J. E. Matthew. *The Colonisation of Greater Britain . . .* Melbourne and London: Chaffey and Gordon and Gotch, 1887.

Vital, David. *The Origins of Zionism.* Oxford: Clarendon, 1975.

Von Geldern. "Reminiscences of the Pioneer Engineers of California." *Western Construction News* 4, no. 2 (1929): 494–95, 555–56.

Voullaire, Kaye. *Mildura Irrigation Settlement: The Early Years.* Mildura: Sunraysia Daily, 1985.

Wadsworth, H. A. "A Historical Summary of Irrigation in Hawaii." *Hawaiian Planters' Record* 37, no. 3 (1933): 124–62.

Wagner, Henry R. "George Davidson, Geographer of the Coast of America." *California Historical Quarterly* 11 (December 1932): 299–320.

Walker, David H. *Ontario, California, "The City That Charms."* San Francisco: Sunset Magazine Homeseekers' Bureau, 1912.

Walker, John Brisben. "The 1915 Exposition and Education." *Sunset* 28, no. 6 (1912): 751–58.

Walsh, Thomas F. "Humanitarian Aspect of National Irrigation." *Forestry and Irrigation* 8 (December 1902): 505–9.

Walter, R. F., and W. H. Code. "Elwood Mead." *American Society of Civil Engineers Transactions* 102 (1937): 1611–18.

Walton, John. *Western Times and Water Wars: State, Culture, and Rebellion in California.* Berkeley: University of California Press, 1992.

Weinmann, H. *Agricultural Research and Development in Southern Rhodesia under the Rule of the British South Africa Company, 1890–1923.* Salisbury: University of Rhodesia, 1972.

Wells, Sydney. *The Chaffey Era: Mildura and Renmark: The Irrigation Colonies.* N.p., 1994.

———. *Paddle Steamers to Cornucopia: The Renmark-Mildura Experiment of 1887.* Renmark, Australia: Murray Pioneer, 1987.

Wescoat, James L., Jr. "Wittfogel East and West: Changing Perspectives on Water Development in South Asia and the United States, 1670–2000." In *Cultural Encounters with the Environment: Enduring and Evolving Geographic Themes,* ed. Alexander Murphy and Douglas Johnson, 109–32. Lanham, Md.: Rowman and Littlefield, 2000.

West, George P. "Riveted Down—And They Like It." *Collier's*, July 29, 1922.

Wheatcroft, Geoffrey. *The Randlords*. London: Weidenfeld and Nicolson, 1985.

Whitcombe, Elizabeth. *Agrarian Conditions in Northern India*. Vol. 1, *The United Provinces under British Rule, 1860–1900*. Berkeley: University of California Press, 1972.

Widney, Robert M. *Ontario: Its History, Description, and Resources . . . Valuable Information for Those Seeking Homes in Southern California*. Riverside, Calif.: Press and Horticulturist, 1884.

Wilcox, Carol. *Sugar Water: Hawaii's Plantation Ditches*. Honolulu: University of Hawaii Press, 1996.

Wilkins, Mira. *The Emergence of Multinational Enterprise: American Business abroad from the Colonial Era to 1914*. Cambridge: Harvard University Press, 1970.

Williams, Alpheus F. *Some Dreams Come True*. Cape Town: Rustica, 1948.

Williams, James C. *Energy and the Making of Modern California*. Akron, Ohio: University of Akron Press, 1997.

Wilson, Herbert M. "American Irrigation Engineering." *Transactions of the American Society of Civil Engineers* 25 (August 1891): 161–222.

———. "Irrigation in India." *Transactions of the American Society of Civil Engineers* 23, no. 454 (1890): 217–53.

———. "Irrigation in India." In U.S. Geological Survey, *Twelfth Annual Report, 1890–91*, 363–561. Washington, D.C.: U.S. Government Printing Office, 1891.

Wilson, James. "Irrigation Creates Home Markets." *Forestry and Irrigation* 8, no. 1 (1902): 10–11.

Wittfogel, Karl. *Oriental Despotism: A Comparative Study of Total Power*. New Haven: Yale University Press, 1957.

"The Witwatersrand Gold-Field and Its Workings." *Engineering and Mining Journal*, July 10, 1897, 271–73.

Woehlke, Walter V. "Food First: How One Western State Is Staking the Farmer." *Sunset* 45 (1920): 9–36.

Wolf, Eric. *Europe and the People without History*. Berkeley: University of California Press, 1982.

Wolpert, Stanley A. *A New History of India*. 4th ed. New York: Oxford University Press, 1993.

Worster, Donald. *Nature's Economy: A History of Ecological Ideas*. 2nd ed. New York: Cambridge University Press, 1994.

———. *Rivers of Empire: Water, Aridity, and the Growth of the American West*. Oxford: Oxford University Press, 1985.

Wright, Hamilton. "How the Commercial Organizations of California Cooperate with the Movement for Government Reclamation." *Forestry and Irrigation* 11, no. 8 (1905): 387–91.

Bulawayo, Rhodesia, 122
Bureaucracy: Progressive belief in, 59, 80, 185; water projects and, 30–31, 48
Burns, Robert, 75
Burton, Rev. A. R. E., 76

Cahuilla Indians, 60, 63
Calexico, Calif., 61, 62
California: agricultural expansion (1850–70) in, 20–21; agricultural trends (1890s–1900s) in, 57–58; anti-Asian sentiment in, 89; Australian Royal Commission visit to, 71–72; British India parallels with, 22–23, 30, 34, 168; British mine financing and, 99; drainage and, 43, 45–46, 47, 49; elite settlers program for, 90; engineering and (see California engineers); farmer-miner conflict in, 35–36; faulty land and water laws of, 26, 27–28 (see also Land policy; Water policy); federal irrigation projects and, 58; frontier development in, 18–19, 20, 37, 39, 40–51, 63, 66, 98, 101; Hawaiian Islands development and, 133–34, 135, 145, 148, 150; hydraulic mining controversy and, 35–36, 42, 43, 46, 49–50, 97, 98–99, 111; hydrographic survey of, 43–44; irrigation projects and (see under Irrigation); lack of government mining regulation in, 18–19; lack of statewide irrigation model for, 39; population growth of, 21; racial hierarchy in, 18, 19, 88–89, 154; scientific colonization and, 90–95; South Africa compared with, 97–98, 114, 126, 127, 168; Victoria compared with, 66, 67–68, 70, 71, 85. See also Gold rush, California; Great Valley; Imperial Valley; Northern California; Southern California; specific place names
California, Gulf of, 60
California Academy of Sciences, 24
California and Hawaiian Sugar Refining Corporation, 136, 149
California Development Company, 58, 60–61, 62–63
California Doctrine (1884), 50
California engineers, 9–14, 179–87; "classic engineer" profile and, 181–82; common themes of, 15, 129, 178; creation of state

office of, 36–37, 39; Exploration Company and, 98–101; export of expertise of, 9–14, 37, 39–40, 66, 95, 97, 111, 161, 179, 180–81, 184 (see also Australia; India; Palestine; South Africa); federal San Joaquin irrigation survey and, 24–25; frontier development ideals of, 101–2, 111; irrigation models of, 39–40, 67; mid-nineteenth-century opportunities for, 41; as "missionaries of civilization," 12, 15; Panama Canal and, 2; personal interest and, 186–87; progress ideal of, 8–9, 10, 13, 14, 15, 16, 66, 115–16, 119, 129, 179, 181, 185; romance of, 78; Social Darwinism and, 6, 7, 8, 15, 185, 186; state office training of, 51. See also Chaffey, Ben; Chaffey, George; Hall, William Hammond; Mead, Elwood
California Farmers' Union, 25
California Land Settlement Board, 14, 90, 92, 93; abolishment of, 94; Mead chairmanship of, 133
California Powder Works, 100
California State Agricultural Society, 25
California State Assembly, 35–36, 45
California State Commission on Colonization and Rural Credits, 89
California State Engineering Office, 36–37, 39, 58, 111, 183; accomplishments of, 51; political controversies and, 42, 43–44, 49, 50–51; weakness of, 47. See also Hall, William Hammond
California State Irrigation and River Commission, 41
California State Rivers and Water Supply Commission, 14
California State Senate, 42
California Steam Navigation Company, 21
California Sugar Refinery, 135–36, 149
California Supreme Court, 50
Campbell, Sir John, 175
Canaan, 78
Canada, 130
Canals: British India and, 12–13, 20, 28, 31–32, 161; Imperial Valley reclamation and, 61, 62, 63; Rhodesia and, 122. See also Irrigation
Cape Colony, 120–28; British control of, 100, 101, 103, 120, 122, 127; California compared

with, 126–27, 168; irrigation plan for, 97, 111, 127–28; land availability and, 127
Cape Town, 121
Capitalism, 1, 12, 13, 15, 39, 40, 107, 158, 160; California engineers' belief in, 98, 111, 180, 183; early expression and rise of, 2, 31; imperialism and, 4; mid-nineteenth-century changes and, 2; Palestine Jewish settlement and, 164–65, 167, 169, 174, 175, 176, 178; progress linked with, 179; as rural development basis, 162, 163, 169; San Francisco world's fair exhibits of, 1; South African development and, 12, 115, 116, 117. See also Laissez-faire culture; Private enterprise
Carmichael (Calif.) settlement, 89
Castle and Cooke (company), 139
Cattle ranching, 20, 21, 28, 34, 58; fencing laws and, 25
C. Brewer and Company, 137, 139
Cement pipelines, 53
Central Valley (Calif.). See Great Valley
Chaffey, Ben (William Benjamin), 14, 52–57, 89; California irrigation colonies and, 52–57, 63, 85, 93, 166, 174; comparison with biblical Moses of, 78; social ideals of, 57, 74, 78, 174, 185; Victoria colony and, 56, 57, 68, 73, 74–81, 82, 180, 183
Chaffey, George, 14, 51, 53–57, 89; California irrigation colonies and, 53–57, 63, 73, 93, 166, 174; comparison with biblical Moses of, 78; Imperial Valley irrigation and, 60–61, 62; ruined Australian career of, 80–81, 82; social ideals and, 57, 74, 78, 174, 185; Victoria colonies and, 56, 57, 68, 71, 74–82, 180, 183
Chaffey, George, Sr., 52
Chaffey Agricultural College, 57
Chaffey Brothers Limited Company, 74, 78, 80
Chamberlain, Joseph, 117, 120
Chapman, William S., 22, 26, 27
Chicago World Columbian Exposition (1893), 1
Chico, Calif., 90
Chico Record (newspaper), 94
China, 7, 29, 30, 184
Chinese immigrants, 19, 21; California agriculture and, 88, 125; California gold mines and, 103; California Vigilantes and, 36; Hawai-

ian Islands and, 144, 158; Ontario (Calif.) colony and, 55, 57; South African mines and, 103, 130; U.S. entry restriction for, 149
Chowchilla (Calif.) settlement, 89
Christian missionaries, 19, 134–35, 137, 138
Citrus crops, 52, 54, 78, 172–73. See also Orange groves
City of San Francisco (steamer), 134
Civic independence ideal, 12, 13, 40, 111, 114–25, 180
Civil engineering, 2, 5, 8. See also Hydraulic engineering; Engineering
Civil War, American, 41, 123, 135; military engineering and, 24
Claus Spreckels (schooner), 141
Clear Lake (Calif.), 59, 60
Closer settlement program (Victoria), 82–87, 89, 155, 175–76, 180; land distribution and, 85–86; limited success of, 86. See also Mead, Elwood
Coal mines, 103
Coastal Range, 21, 22, 90
Coast and Geodetic Survey, U.S., 7, 24
Cobden, Richard, 180
Collectivism, 162, 163, 166, 170, 175; Joint Palestine Survey Commission critique of, 177
Collier's (magazine), 91
Colonialism. See Imperialism
Colorado, 14, 71, 99
Colorado Desert, 60–63
Colorado River, 59–62; flooding of, 62, 64–65
Colorado School of Mines, 182
Colorado State Agricultural College, 83
Columbia River, 5
Committee for the Relief of Belgium, 184
Common law, 42, 138
Community rights, 19, 52, 53
Computers, 15
Comstock Lode, 18, 21
Congress, U.S.: federal reclamation and, 59; foreign irrigation system inquiries by, 29; Hawaiian Homes Commission and, 133; irrigation policy and, 23–24, 34
Conservation, 87. See also Environment
Consolidated Gold Fields. See Gold Fields Consolidated

Contract labor, 134, 144, 145, 148, 149

Cook, James, 134, 136

Cooke, George P., 133, 150, 155

Coolgardie gold fields, 184

Cooperative agriculture: Delhi (Calif.) and, 93; Jewish Palestine and, 162, 165, 167, 173–74; Mildura (Victoria) and, 78; Ontario (Calif.) and, 51, 52, 53–54, 58

Corner House, 114, 116

Corn Laws repeal of 1846 (Britain), 20

Corporations: California landholding by, 88; cooperative fruit farming and, 58; hydraulic mining and, 35; laissez-faire policy and, 18–19; mutual water company and, 52

Corps of Engineers. *See* Army Corps of Engineers, U.S.

Corvee labor, 22, 31, 33

Craig, Hugh, 135, 158

Cucamonga Mountains, 53, 54

Cummins, Thomas, 137

Cunningham, Sir Edward, 71

Curti, Merle, 7

Dams: British India and, 29, 31, 32, 161; federal projects and, 61; hydraulic mining and, 18, 46–47, 49, 111, 140; irrigation canals and, 26, 31; Johannesburg water supply and, 111

Darwin, Charles, 6, 76

Davidson, George, 18, 25–26, 27–36, 43, 47; background of, 24; British India hydraulic projects and, 31–33, 161, 168, 184; Drainage Commission and, 45; *Irrigation and Reclamation of Land for Agricultural Purposes, as Now Practiced in India, Egypt, Italy, etc.*, 34; as planned irrigation advocate, 35–36, 48, 127; riparian principle and, 27, 42, 50

Davies and Company, 139

Dead Sea, 164, 168

Deakin, Alfred, 19, 28, 32, 54, 80, 124, 180; Australian irrigation system and, 69, 70–71, 73, 82, 87, 127; background of, 70; Chaffeys' contract with, 74; *Irrigation in Western America*, 72; Mildura model colony and, 74, 81; political beliefs of, 80, 84, 165; on Victoria-California parallels, 67, 68, 71; visit to America of, 70–72; water policy of, 73

De Beers Mining Company, 100, 101, 103, 124, 125

Debris Commission, 47

Debris dams, 46–47, 49, 111, 140

Debris problem, 35, 36, 40, 42–44, 50

Declaration of Independence (U.S.), 118–19

De Crano, Edmund, 98–99, 100, 101, 107

Deep-level mining, 107, 110, 185

Deforestation, 70

Delagoa Bay, 103

Delhi (Calif.) model colony, 92–94, 154, 155–56, 157, 159, 168, 169, 171. *See also* Chaffey, Ben; Chaffey, George

Democratic society, 28, 39, 51, 130, 166; California engineers' attempted transfer of, 98; closure of frontier and, 93. *See also* Civic independence ideal

Denmark, 89; workers from, 149

Depression of 1870s, 23, 27

Depression of 1893, 50, 58, 59, 78, 80, 81, 82

Depression of 1930s, 86, 94

Desert Lands Act of 1877 (U.S.), 20

Deuel, C. H., 94

Diamond mines, South Africa, 100, 101, 103, 127

Díaz, Porfirio, 62

Disease, 141, 152, 178, 180

Disraeli, Benjamin, 99

Doornfontein reservoir, 110

Dow, John L., 71

Downey, John G., 42

Drainage Act of 1880 (Calif.), 45–46; repeal of, 47, 49

Dreyfus case (France), 163

Drought: Australia, 68, 72, 75; British India, 30; California, 21, 36, 40, 49, 50, 58, 59; Transvaal, 110, 111

Du Bois, W. E. B., 8

Durham (Calif.) model colony, 90–92, 93, 94, 154, 156, 157, 159, 168, 169, 171. *See also* Chaffey, Ben; Chaffey, George

Eads, James B., 45, 46

East India Company, 165

East India Railway, 32

East Maui Irrigation Company, 137

East Rand coal field, 103
Ecological planning. *See* Environment
Economic depressions. *See* Depression *headings*
Economic growth: Californian ideals and, 39–40, 63; government's role in, 37, 39, 58, 59, 80, 89; Mead's views on, 174, 177. *See also* Global economy
Eden, 56
Education: agricultural, 57, 78; engineering, 83, 84, 118, 182. *See also* University of California, Berkeley
*Edwards Woodruff v. North Bloomfield Gravel Mining Co., et al* (1884, Calif.), 50
Egypt, 29, 30, 172
Eiffel Tower, 1
Eighteenth Amendment, repeal of, 57
Eilat, Gulf of, 163
Electric lighting, 54, 90, 140
Elitism. *See* Social hierarchy
El Niño, 62
Ely, Richard T., 8
Emek. *See* Esdraelon Plain
Engels, Friedrich, 2
*Engineer, The* (Hammond), 12, 16, 182, 186
Engineering, 181–87; American prominence in, 99–100; evolving prestige of, 183–84; ideas about race and, 186; method and, 186; professional education and skills of, 182; progress ideal and, 8–9, 14, 15, 16, 179, 185; shift in ideology of, 185–86; Social Darwinism and, 15, 185. *See also* California engineers; Hydraulic engineering; Mining engineering
*Engineers and the Price System, The* (Veblen), 12, 182
England. *See* Great Britain
English common law, 42, 138
English Dam (Calif.), 18
Environment, 12, 15, 181; agricultural runoff damage to, 33, 42; damage in British India to, 33; hydraulic mining's impact on, 35, 42, 43–44, 46; state resources planning and, 47; technological conquest of, 2, 77, 78
Equality concept, 15. *See also* Racial hierarchy; Social hierarchy
Esdraelon Plain, 163, 164, 166, 168
Ethnic cleansing, 15

Ethnic hierarchy. *See* Racial hierarchy; Social hierarchy
Etiwanda (Native American chief), 53
Etiwanda colony (Calif.), 51, 52, 53–54
Etiwanda Water Company, 54
Eucalyptus, 168
Euclid Avenue (Ontario colony), 54, 75
Eugenics, 6
Exclusion Acts of 1880s (U.S.), 149
Exploration Company, 98–101, 110, 111, 130
Exports. *See* Trade

Famine: British India irrigation and, 12–13, 22, 30, 31
Farmers. *See* Agriculture; Yeoman farmer ideal
Farrar, George, 115, 116, 117
Farrington, Wallace R., 152, 156
Feather River, 18, 30; mining debris silting of, 35, 43–44
Federal government. *See* Government
Ferris wheel, 1
Feudalism, 31
Floods, California, 35, 36, 43, 44, 45, 50, 62, 64–65; preventive measures, 46
Fluid mechanics, 2
Fontana (Calif.) settlement, 89
Food Administration, U.S., 184
Ford car, 1
Foreign Miners Tax (Calif.), 19
Fossil fuels, 2
France, 4, 100; irrigation system, 29, 48, 49
Frankel, Lee K., 170–71
Free markets. *See* Capitalism
Fresno, Calif., 59, 72
Fresno Slough, 22
Friedlander, Isaac, 22, 26, 27
Friedman, Thomas, 15
Frontier development, 10; America's closing of, 93; Australia and, 75–76; California and, 18, 37, 39, 40–51, 63, 66, 98, 101; engineers' role in, 41, 101–2, 111, 130; Hall's vision of, 120–23; model irrigation colonies and, 51–58; romanticized image of, 75–76; social and political issues and, 183; South Africa and, 97, 98, 121, 130
Fruit farming: California and, 52, 54, 58, 59, 90,

and, 17, 21–28, 36–37, 42–43, 58; Japanese
  farmers in, 88–89
Growers' Cooperative Association, 54
Grunsky, Carl E., 51, 62
Gum trees, 75

Hackfield and Company, 139
Haggin, James Ben Ali, 22, 31, 50
Haifa, Palestine, 168, 172
Haiku Ditch (Hawaii), 140
Haleakala Mountain (Hawaii), 138
Hall, William Hammond, 13–14, 19, 40–51, 62,
  150; accomplishments of, 51; background of,
  40–41; as California state engineer, 36–37, 39,
  40, 41, 43–51, 111, 183; California state irrigation district proposal of, 48–49, 50, 63; contentious personality of, 111; debris dams and,
  46–47, 140; frontier development and, 120–
  23; investigation/acquittal of, 51; problems
  of, 47–48, 49, 58; professional reputation of,
  42; progress ideal of, 115–16, 119, 129; racial
  and social beliefs of, 123–24, 126, 127–28, 129;
  South African development and, 95, 98, 101,
  111, 115–16, 120–30, 155, 168, 182–83; Victoria
  settlement and, 70–73, 180; water law and,
  49, 184–85
Hamakua Ditch (Hawaii), 137
Hammond, John Hays, 9, 12, 14, 16, 40, 95,
  181, 186; American ideals of, 118–19, 129,
  130; deep-level mining and, 107, 110, 185;
  The Engineer, 12, 16, 182, 186; on engineering profession, 181–83, 185, 187; high treason
  sentence of, 119; Jameson Raid and, 116–18;
  mining background of, 118; South African
  development and, 97, 101, 103, 107, 110, 111,
  114–19, 125, 130, 180, 182, 183
Hammond, Natalie Harris, 119
Hammond, Richard Pindell, Jr., 40
Harding, Warren G., 152
Hard rock mining, 97
Harriman, E. H., 62–63
Hawaii (island), 153, 156, 159, 169
Hawaiian Commercial and Sugar Company,
  139, 141, 149–50
Hawaiian Home Lands Department, 159
Hawaiian Homes Commission (HHC), 133,

150–60, 169; homesteader problems and,
  156–57, 158; ideals of, 174; interracial mar-
  riage and, 159; land policy and, 166
Hawaiian Islands, 133–60; ancient land laws
  of, 137; Asian immigrant influx into, 158; beginning of American interests in, 134–35;
  California engineering transfer to, 10, 11, 12,
  13, 14, 40, 134; climate of, 134; irrigation system and, 29, 133, 134, 137–40, 148, 150, 155–57;
  native population of (*see* Hawaiian natives);
  plantation labor shortages and, 145; progress
  concept and, 130–31, 159; racial intermarriage
  and, 158–59; single-industry dependence of
  (*see* Sugar industry); size and land surface of,
  134; U.S. annexation of, 136, 151, 152, 154
Hawaiian Land Act of 1895 (U.S.), 152
Hawaiian natives, 144; controversial definition of, 151, 153, 186; disinterest in profit, 158;
  homesteading resettlement program for,
  95, 131, 133–34, 150–60, 161, 169, 174, 180, 186;
  population decline of, 141, 144, 152; racial
  intermarriage of, 158–59. *See also* Hawaiian
  Homes Commission
Hawaii Organic Act of 1900 (U.S.), 152
Haymond, Creed, 36, 42
Hays, John, 40
Heber, Anthony, 62
Hebrew University of Jerusalem, 168
*Helping Men Own Farms* (Mead), 150
Henriques, Cyril Q., 171
Henshilwood, James, 80
Herzl, Theodor, 165
HHC. *See* Hawaiian Homes Commission
HHC Act of 1920 (U.S.), 151, 152–53, 154
Hilo, Hawaii, 152
Himalayas, 31, 32
Histadrut, 174
Hobsbawm, Eric, 4–5
Homestead Act of 1862 (U.S.), 20, 69, 152
Homesteading. *See* Model irrigation colonies;
  Rural life
Honolulu, 152, 157
Honolulu Social Science Association, 152
*Honolulu Star-Bulletin*, 154
Honolulu Water Works, 150
Honouliuli Rancho (Hawaii), 150

Hoover, Herbert, 7, 183–84
Horticulture. *See* Agriculture
Hula Basin (Palestine), 178
Hula Marshes (Palestine), 168
Humboldt University, 168
Hyde Park (London), 1
Hydraulic engineering, 2, 9; British India and, 12–13, 17, 20, 22, 27, 28–37, 39, 40, 45, 48, 70, 161, 184; California State Office training and, 51; Coast and Geodetic Survey and, 24; Palestine/Israel and, 177–78; personality and skill set for, 181–82. *See also* Dams; Engineering; Irrigation
Hydraulic Miners' Association, 50
Hydraulic mining, 40, 42–46, 99; California gold rush and, 18–19, 98–99; debris and damage from, 35–36, 42, 43, 46, 111; lack of government regulation of, 18–19; landmark case against, 49–50, 97
Hydraulic society thesis, 30–31
Hydroelectric power, 54, 56, 61, 78
Hydrographic surveys, 43–44

Idaho mine financing, 99
Immigration Act (Australia), 70
Immigration Bureau (Hawaii), 144
Imperial Irrigation District (Calif.), 63
Imperialism, 10, 13, 100, 120; Britain and, 4, 97–98, 100, 101, 117, 122, 180; Hawaiian economy and, 133, 136; inequalities and, 4–5, 33, 95, 121, 123, 129, 175, 181; infrastructure projects and, 31, 33; progress ideal and, 5–6, 15, 30, 119, 179; scientific and technological aids to, 180; settlement model of, 162; United States and, 4, 136, 151, 152, 154, 180; "white great mission" of, 40, 179. *See also* India; South Africa
Imperial Land Company (Calif.), 61
Imperial Valley (Calif.), 58, 60–63, 86, 89, 168; flooding of, 62, 64–65; land policy and, 61
India: British government's hydraulic program in, 12–13, 17, 20, 22, 27, 28–33, 34, 39, 40, 45, 47, 48, 70, 73, 161, 166, 184; California parallels with, 22–23, 30, 34, 168; Davidson's visit to, 31–33, 34; flaw in hydraulic regime of, 33; South Africa and, 100

Indian Peninsula Railroad, 22
Individualism, 93
Indus River, 30, 32
Indus River Delta, 31, 33
Industrial capitalism. *See* Capitalism
Industrial Commission of Inquiry (Johannesburg), 115
Inferior races. *See* Racial hierarchy
Infrastructure, 10, 31, 33
Institute of Agriculture and Natural History, 172
Interior Department, U.S., 59, 63, 86. *See also* Reclamation Service, U.S.
International fairs. *See* World's fairs
Internet, 15
Inyanga (Rhodesia), 122
Irish workers, 149
Irrigation: adverse environmental effects from, 33; ambiguous water laws and, 19–20; Australian projects and, 40, 68–73, 78, 82–95, 161; British India system of, 22, 28–33, 48, 73, 161; California's models of, 10, 11, 13, 14, 17–18, 19, 20, 21–28, 39–43, 51–63, 66–70, 73, 168; California's overall pattern of, 58, 66; California's proposed statewide system of, 26; California State Engineering Office and, 36–37, 41, 42, 43, 48–49, 50; California's vested interests and, 47; centralized government power and, 23–24, 30–31, 73; Chaffey brothers' technical innovations and, 53; creation of districts for, 69; Davidson's international study of, 28–36; first U.S. engineering professor of, 83; Hawaiian Islands and, 29, 133, 134, 137–40, 148, 150, 155–57; hydraulic society thesis and, 30–31; international models of, 39–40; Jewish Palestine and, 29, 166, 168–69, 170, 177, 178; local control of, 39, 40, 66, 73; Mead's expertise in, 83–84, 93; mining debris and, 43; mountain snow-melt source of, 26, 71, 75; pipelines and, 53, 79; South Africa and, 29, 97, 111, 115–16, 121–22, 126–30, 186; U.S. government projects and, 58–66. *See also* Closer settlement program; Model irrigation colonies
Irrigation Act of 1886 (Victoria), 73, 74
*Irrigation Age* (journal), 182
Irrigation Department (British India), 32

Irrigation Investigation Office, U.S., 84

*Irrigation in Western America* (Deakin), 72

Irwin, Will (California governor), 36, 37, 41, 42

Irwin and Company, 134, 139, 148–49

Islam, 170

Isolationism, 7

Israel, state of, 170, 177, 178

Isthmian Canal Commission, 5

Italy, 4, 29, 48, 49

Jaffa, Palestine, 163, 167

Jaffa orange, 172–73

Jameson, Leander Starr, 97, 116, 117, 118

Jameson Raid (1895), 97, 114, 116–18, 119–20, 124, 183

James Robinson and Company, 137

Japan, 4, 29; California farmers from, 88–89; Hawaiian immigrants from, 140, 144, 145, 154, 157, 158, 159

Jefferson, Thomas, 39, 87–88, 89

Jerusalem, 161, 167

Jewish Agency, 177

Jewish Colonization Association, 164–65

Jewish Federation of Labor, 173, 174–75, 176; critique of, 177

Jewish National Fund, 165, 169, 170, 176–77; settler model contract with, 176

Jewish national homeland, 13, 165, 169, 175. *See also* Zionist movement

Jewish settlers. *See* Palestine

Jezreel Valley (Palestine), 164, 168

Johannesburg, South Africa, 14, 97, 110, 111, 121, 122; black labor and, 103; British takeover of, 130; gold rush influx into, 102; Jameson Raid and, 116–18

Johannesburg Waterworks Estates, 110

Johnson, Hiram V., 86, 89–90

Joint Palestine Survey Commission, 170–78

Jordan River, 164, 165, 168, 178

Jordan Valley, 168, 177, 178

Jumna River, 30, 32

Justinian law, 19

Kahuku Rancho (Hawaii), 150

Kahului, Hawaii (Maui port), 141

Kahului Railroad, 141

Kalakaua (king of Hawaii), 130, 136, 138, 149

Kalaniana'ole, Jonah Kuhio (prince of Hawaii), 152

Kalanianaole settlement (Molokai), 153–55, 156–57

Kamehameha III (king of Hawaii), 137

Karoo (South Africa), 121

Kathedersozialisten, 8

Kauai, 136

Kaunakakai, 157

Kern River, 19, 23, 50

Kibbutz (modern movement), 170, 178. See also *Kvutzah*

Kimberley, South Africa, 101, 103, 121

King's River, 59, 60

Kishon Valley (Palestine), 169

Klip River (South Africa), 111

Koloa, Hawaii (Kauai island), 136

Koolau Ditch, 150

Kopper Kerk sect, 115

Korean workers, 144, 157

Kreutzer, George C., 90, 91

Kruger, Paul, 100, 110, 111, 114–15, 116; abdication and exile of, 130; conservatism of, 129; results of failed Jameson Raid and, 117, 119, 120

*Kvutzah* (Jewish settlements), 165–66, 169, 174; subsidization of, 175. *See also* Kibbutz

Labor sources: British India corvee policy and, 22, 31, 33; California agriculture and, 20, 21, 55–56, 57, 88, 123; Hawaiian Islands and, 133, 134, 140, 141, 144–45, 148, 149, 158; Palestine Jewish settlements and, 163, 164, 167, 173, 174–75, 176, 177; self-labor ideal and, 154, 158, 167, 175; social and racial hierarchy in, 57, 88, 90, 124, 125–26, 144–45, 158; South Africa and, 98, 100, 103, 119, 123, 126, 127, 130; squalid conditions and, 145; Transvaal's unjust system of, 115–16

Ladd and Company, 136

Laguna Weir, 61

Laissez-faire culture, 180; American ideals and, 114; California's development and, 18–19, 37, 40, 49, 66; hydraulic mining damage and, 35. *See also* Capitalism

Lanai (Hawaiian island), 138

83–84; mistakes of, 94; Palestine Jewish agricultural development and, 95, 161, 163–64, 167–77, 178, 180, 185; racial views of, 154, 158; scientific colonization ideal of, 90–95, 155–56, 157, 172, 174; self-labor belief of, 154, 158; state-settlement scheme of, 94–95; U.S. Reclamation Commission and, 171; Victoria settlements and, 68, 71, 82, 83–87, 89, 155; water management models and, 178. *See also* Closer settlement program

Mediterranean Sea, 164

Melbourne, Australia, 80, 84, 85

*Melbourne Age*, 86

*Melbourne Argus*, 71, 76

Mendell, George, 24, 36, 43–44, 49

Merced, Calif., 72, 92

Mesopotamian hydraulic society, 30

Mexicali, Calif., 61, 62

Mexicans: Anglo Californian settlements and, 18, 19, 55, 114; as California agricultural workers, 57, 88, 125; California ranchos and, 19

Mexican War, 41

Mexico, 29, 99, 118; Imperial Valley development and, 61, 62, 63

Migrant workers, 57, 148

Mildura Irrigation Company, 75, 89

Mildura model colony (Victoria), 73, 74, 75–82; economic problems of, 78, 80–82, 92; renewed life of, 81–82; settler's backgrounds and, 76–77; social ideals of, 185. *See also* Chaffey, Ben; Chaffey, George

Mildura Settlers' Club, 78

Mildura Winery, 82

Military engineering, 24

Mill, John Stuart, 76

Miller, Henry, 22, 23, 26, 34, 41; riparian rights case and, 50

Milner, Lord Alfred, 120, 130

Miners' Association, 35, 99

*Mining and Scientific Press*, 103

Mining engineering, 7, 10, 11, 13, 37; Australia and, 40, 70; California and, 17, 18–20; deep-level mining, 107, 110, 185; farmers' conflict with, 11, 35–36, 99; Hoover and, 184; land and water rights and, 19; personality and skill set

for, 181–82, 183; South Africa and, 11, 12, 14, 97, 100–101, 107, 114, 121, 122, 130, 186. *See also* Engineering; Gold rush *headings*; Hydraulic mining

Mining securities, 99

Model irrigation colonies, 39, 51–58, 63, 68, 82–87, 89–94, 180; defined principles of colonization and, 171–72; failures of, 93–94; Hawaiian native resettlement program and, 133–34, 150–60, 169; hierarchical labor system and, 125–26; as Jewish Palestine settlements' model, 168, 169; land policies and, 166; social ideals and, 185; South African attempt at, 125, 127, 130; temperance and, 78, 125, 174, 180; Victoria and, 73, 74–82 (*see also* Closer settlement program); Victorian improvement on California colonies, 85, 90. *See specific names of colonies*

Molokai experiment (Hawaii), 133–34, 150–60, 169; obstacles of, 159–60

Monopolies: hydraulic mining industry and, 35–36; land grants and, 26, 34, 40, 58, 67; social and agrarian reform and, 87; sugar production and, 149; water rights and, 21, 23, 25, 27, 40, 48, 139

Moore, Charles C., 5

Morison, George S., 179, 180

Mormons, 51–52, 71, 140

Mosely, Alfred, 124, 127

Moses (biblical), 78

Moshavim (Jewish small holders' settlements), 169, 174

Mozambique, 103

Mud deposits, 35

Mulgoa scheme (New South Wales), 80

Mumford, Lewis, 8, 183

Municipal services, 18, 56

Murray, Stuart, 69–70, 80, 81, 82, 83

Murray River (Australia), 69–70, 71, 73, 74, 161; federal government's use division of, 83; origination and course of, 75; pumping up of, 78

Murrumbidgee River (Australia), 161

Mussel Slough (Calif.), 19

Mutual water companies, 52, 53–54, 56, 61, 75

Myers, Francis, 76, 77

134, 141, 149; imperialist construction of, 31;
India and, 22, 31; as progress embodiment,
179; refrigerated cars and, 58; South Africa
and, 103, 121, 122, 125, 139–40; steel design of,
5. *See also specific lines*

Rainfall: California and, 21, 49, 60, 62; Hawaiian Islands and, 134; Palestine and, 164; Punjab and, 32; South Africa and, 121; Victoria
and, 69, 71, 75; Western Europe and, 31

Ralston, William C., 18, 21–22, 23, 26, 27–28,
36–37; bankruptcy of, 27, 36; death by
drowning of, 27, 28

Ranchers. *See* Cattle ranching

Rand. *See* Witwatersrand

Rand Club, 101

Reciprocity Treaty of 1876 (United States–
Hawaii), 134, 135, 136, 144

Reclamation Act of 1902 (U.S.), 59, 83

Reclamation Bureau, U.S., 14, 23, 95, 177

Reclamation Service, U.S., 8, 59–63, 84, 86, 95,
145, 150, 171, 183; as Australian influence, 83;
creation of, 59; private interests' battle with,
58, 61–62; weaknesses of, 63

Red Bluff, Calif., 21, 27

Red Book (Mildura colony advertising), 75–76,
80

Redlands, Calif., 126

Reforestation, 176

Refrigerated transport, 58, 81

Renmark model colony (South Australia), 73,
80

*Report on the Lands of the Arid Region of the
United States* (Powell), 23

Reservoirs, 23, 59, 110, 126; federal projects
and, 61

Rhodes, Cecil, 95, 97, 100–101, 103, 107, 110,
111, 114, 120, 121, 129, 130; Exploration Company alliance with, 101; Jameson Raid and,
116–17; scientific agricultural development
and, 122–25

Rhodes, Frank, 116

Rhodes Fruit Farms Company, 125

Rhodesia, 97, 116, 120, 122–23

Rice production, 88

Rights-of-way, 23

Riparian principle, 22, 23, 27, 42, 138;

British India and, 30; landmark case and, 50;
prior appropriation vs., 19, 42, 50

Risdon Iron Works, 140

Rishon Le-Zion settlement (Palestine), 163

Rivers, 18, 46; mining debris and, 35; navigation
of, 43; riparian law and, 19, 23. *See also* Irrigation; *specific rivers*

Riverside, Calif., 51, 52–53, 126

*Rivers of Empire* (Worster), 31

Robinson Deep, 110, 111

Rockwood, Charles Robinson, 60–61, 62

Rocky Mountains, 60

Rodgers, Daniel, 7–8

Rodney Irrigation Trust, 73

Roman irrigation system, 170

Roosevelt, Theodore, 59, 63, 84

Rosenwald, Julius, 171

Rothschild, Edmond de, 99, 164, 176

Rothschild family, 98, 101

Royal Hawaiian Agricultural Society, 144

Royal School of Mines (Freiburg, Germany),
118, 182

Rudd, Charles, 101, 107, 117

Rule of law, 179

Ruppin, Arthur, 168

Rural credit system, 85, 89, 90, 94

Rural life, 86; advertising of, 75–76; economic
hardship and, 89; encouragement of, 88, 94,
95, 154; idealism of, 165, 185; ideological agendas and, 185; Zionist movement and, 162, 163,
165. *See also* Model irrigation colonies; Yeoman farmer ideal

Ruskin, John, 75

Russia, 4, 163

Ryerson, Knowles A., 171, 172, 173, 177

Sacramento, Calif., 42, 46

Sacramento River, 18, 21, 26, 30, 59; battle over
control of, 35; mining debris silting of, 35,
36, 43

Sacramento Valley, 21, 24, 59; farmer-miner
conflict and, 35, 99; government protective
measures and, 46, 47; hydraulic mining environmental destruction in, 43–44, 46; scientific colonization project and, 90

St. Louis World's Fair (1904), 56

Salinity, 33

Salt Lake City, 52, 71

Salt Lake Railroad, 56

Salton Sea, 62, 63

Salton Sink, 60, 62

San Antonio River, 54

San Bernardino, Calif., 51, 126

San Bernardino Basin, 168

San Bernardino Mountains, 54, 59

San Bernardino Rancho, 51–52

San Bernardino Valley, 53

Sand deposits, 35

Sandwich Islands. *See* Hawaiian Islands

San Francisco, 24, 41, 71–72, 89; gold rush development of, 101; Hawaiian sugar empire and, 131, 134, 135, 136, 140, 141, 148, 149; importance as port of, 18, 21; public utilities and, 56; Ralston irrigation project and, 21–22. *See also* Panama-Pacific International Exposition

San Francisco Bay, 22, 27

San Francisco Chamber of Commerce, 135, 158

*San Francisco Commercial Herald*, 140

*San Francisco Examiner*, 2

San Francisco Park Commission, 41

San Francisco Sugar Refinery, 135

San Francisco Water Works Company, 21

San Francisco World's Fair (1915). *See* Panama-Pacific International Exposition

San Gabriel Mountains, 54, 59

San Joaquin and King's River Canal Company, 17–18, 21, 23–28, 29; Hall and, 36–37, 41; reasons for failure of, 27–28

San Joaquin River, 21, 22, 23, 26, 30, 36; charting of, 43

San Joaquin Valley, 19, 20–21, 22, 23, 24, 25, 49, 168; cattle ranchers and, 28; opposition to drainage bill in, 45; war veterans land settlements in, 92

San Jose-San Francisco Railroad, 18

Santa Ana River, 52, 53

Santa Barbara, Calif., 126

Santa Fe Railroad, 53, 56, 92

Santa Rita Ranch, 22

Sawyer, Lorenzo, 50, 97

Schussler, Herman, 139, 140

Schuyler, James D., 49, 51, 150

Schwartz, Mark, 170, 171

Scientific agriculture, 122–23, 172, 176

*Scientific American* (journal), 28

Scientific colonization, 90–95, 172

Scientific knowledge, 179, 180

Scientific management, 185–86

"Scramble for Africa," 4, 100, 120

Self-labor ideal, 154, 158, 167, 175

Semiarid climate, 11, 22, 30, 31, 32, 60; Australia and, 69, 71; Palestine and, 162, 163; South Africa and, 98

Settlers. *See* Frontier development; Imperialism; Model irrigation colonies; *specific place names*

Sheffield Scientific School, Yale University, 118

Sherman, William Tecumseh, 23

Sierra Madre, 77

Sierra Nevada, 11, 20, 21, 71; hydraulic mining and, 18, 35; snow-melt irrigation from, 26, 71, 75, 90

Silt, 60, 62, 63, 70

Silver, 18, 21

Simmer and Jack mine, 110

Sinai Peninsula, 163

Sind (Indus Valley), 30

Single tax movement, 72–73

Sirhind Canal (India), 32

Sivewright, Sir James, 111, 126

Slovenian workers, 149

Smith, Hamilton, 98–99, 100, 101, 107, 110, 111

Smythe, William E., 52, 61, 62

Snake River, 5

Snow-melt, 26, 71, 75, 90

Snowy Mountains, 75

Soane Canals (India), 32

Social change. *See* Progress; Social reform

Social Darwinism, 6, 7, 15, 185, 186

Social hierarchy, 10, 13; Australian "new beginning" image and, 76; California agriculture and, 88, 90, 125–26; equality concept and, 15; Hawaiian Islands and, 134, 150; hydraulic societies and, 31; imperial inequalities and, 33, 98, 121, 122, 129; Jewish Federation of Labor efforts against, 174–75; Ontario colony elitism and, 51, 55–56; self-labor and, 154;

South Africa and, 95, 101, 124. *See also* Racial hierarchy

Social insurance, 7, 8

Socialism: Jewish Palestine settlements and, 13, 162, 165–66, 167, 169; Jewish private Palestine settlements vs., 169, 171, 174, 176, 177; Victorian state socialism vs., 83

Social mobility, 31, 39, 51; California engineers' attempted transfer of, 98; closure of frontier and, 93; Mead's theories and, 88; Victoria and, 76, 77, 82

Social reform, 8, 9, 14, 82; Chaffey brothers' colonies and, 57, 74, 78, 174, 185; Henry George and, 72–73; Jewish Federation of Labor colonies and, 173, 174–75, 176; scientific solutions and, 186. *See also* Model irrigation colonies; Temperance

Social Science Club (Honolulu), 145

Soil Conservation Service, U.S., 177

Soskin, Selig E. H., 167–68

South Africa, 97–130; California compared with, 97–98, 114, 126, 127, 168; California engineering transfers to, 10, 11, 12, 13, 14, 40, 95, 97–98, 100–101, 111, 114–25, 130, 155, 180, 181, 182–83; colonization system for, 89; economic conditions in, 98; English settlers and, 127; frontier society and, 97, 98, 121, 130; government-aided economic growth and, 58; hostility to American ideals in, 130; irrigation and, 29, 97, 111, 115–16, 121–22, 126–30, 186; mining vs. agriculture in, 130; racial hierarchy and, 95, 98, 101, 103, 119, 123, 124–26, 154, 186; wealth sources of, 121. *See also* Cape Colony; Mining engineering; Transvaal

South America, 4, 7

South Australia, 69, 75; model irrigation colony, 80, 95

Southern California, 49; Australian Royal Commission visit to, 71; Cape Colony compared with, 126–27, 168; Drainage Act and, 46; federal reclamation projects and, 59–60; land boom and, 51, 54–57; model irrigation colonies and, 39, 51–57, 58, 68, 85, 125, 127; Palestine as replica of, 168, 172; urban growth of, 58

Southern Pacific Railroad: Imperial Valley and,

60, 61, 62, 63; land and water rights and, 26; land settlements and, 90, 92; Los Angeles and, 52, 53; Ontario colony and, 54, 56; San Francisco and, 149; San Joaquin Valley and, 21

Spain, 29, 48, 49

Spanish-American War (1898), 136

Spanish Law of Waters (1866), 73

Spanish missionaries, 19

Speculation, land, 20, 54–55, 88, 155

Spencer, Herbert, 15

Spreckels, Claus, 131, 133, 134, 135–50, 159; background of, 135; innovations of, 140–41, 148–49; on labor question, 145; land and water rights and, 138–39; peak of power of, 148; sugar-making patents of, 149

Spreckelsville, Maui, Hawaii, 134, 140, 148–49

Spring Valley Water Company, 21, 139

Stanford University, 184

State government. *See* California State *headings*; Government

State socialism (Victoria), 68, 73, 83, 94–95; international socialism vs., 83

States' rights: water development and, 84

Steamboat Slough, 43

Steam power, 77

Steamships, 2, 4, 81, 134, 141, 149; as progress embodiment, 179

Steel, 2, 5

Stewart, William Morris, 23, 24

Stock Breeders' Association, 91

Stock Exchange (Johannesburg), 101

Stockton, Calif., 21, 40

Strahorn, Arthur T., 171

Sub Nigel mines, 110

Suez Canal, 2, 99, 100, 161, 180

Sugar industry, 13, 121, 131, 134, 135–50, 151, 159; labor sources/conditions and, 144–45, 149; mechanized processing and, 140–41; native Hawaiians and, 152, 153; refineries and markets, 141, 149. *See also* Spreckels, Claus

Sunkist oranges, 57

*Sunset Magazine*, 5, 7, 93

Surface mining, 18

Swamp Land Act of 1850 (U.S.), 20

Swedish workers, 149

Swinburne, George, 83
Sydney, Australia, 161
Syria, 29, 172

Taxation, 19, 44, 46; single tax movement and, 72–73
Taylorism, 185–86
Technology: agricultural mechanization and, 20, 57, 77, 162–63, 172, 173, 176; cultural values and, 15; engineers' belief in, 185; foreign transfer of, 10–11, 40, 181; global effects of, 1–2, 4, 5, 7, 9–14, 15, 180; modern farming and, 20, 57, 77, 173, 176; Progressive beliefs and, 59; progress linked with, 179; San Francisco World's Fair exhibits of, 1, 179; South African mining and, 103; sugar production mechanization and, 140–41, 148; turn-of-twentieth-century belief in, 5, 8–9, 14, 16, 39, 77, 78. *See also* Mining engineering; *specific names of engineers*
Tel Aviv, 167
Telegraph cables, 2, 31, 179, 180
Temperance, 57, 72, 78, 125, 174, 180; Mildura settler rejection of, 78. *See also* Social reform
Tenant farmers, 58, 88, 127
Tennessee Valley Authority, 177
Tennyson, Alfred, Lord, 75
Tevis, Lloyd, 22
Texas Rangers, 40
Thurston, Lorrin A., 145
Tolstoy, Leo, 163
Tomato production, 89
Topographical Engineers, 24
Toynbee Hall (London), 8
Trade: Australian agricultural exports and, 81; California agricultural exports and, 20–21, 22, 58, 173; San Francisco port and, 18, 21; United States–Hawaiian Islands, 135, 136
*Transactions of the American Society of Mining Engineers* (journal), 182
Transcontinental railway, 2, 4, 41, 56
Transportation: agricultural market growth and, 20; Cape Colony fruit exports and, 127; Hawaiian native homestead farms and, 157; Hawaiian sugar industry and, 133, 134, 141;

149; mid-nineteenth-century changes in, 2, 4; Mildura colony problems with, 80–81, 82; Palestine Jewish agricultural settlements and, 173; San Francisco network of, 18; San Francisco World's Fair Palace of, 1; shrinking world and, 5–6; South African weakness in, 98. *See also* Railroads; Steamships
Transvaal, 114–25; agricultural potential of, 129; Anglo-Boer War and, 103, 120; Boer biases and, 129; drought and, 110, 111; gold resources of, 97, 100, 101–2, 120; irrigation and, 97, 115–16, 121. *See also* Boers; Witwatersrand
Transvaal Chamber of Mines, 110
Treasury Department, U.S., 29
Tree planting, 176
Triple Canal colonies (India), 32
Tulare Lake, 22, 23, 41
Tulare Valley, 23
Turlock Irrigation District, 92
Turner, Frederick Jackson, 93
Twain, Mark, 99–100, 117, 135, 144, 163

Uganda, 165
Union Iron Works, 100
United States: frontier closure by, 93; Hawaiian annexation by, 136, 151, 152, 154; Hawaiian laborers from, 144; Hawaiian Reciprocity Treaty and, 134, 135, 136, 144; imperialism and, 4, 136, 151, 152, 154, 180; Jewish immigrants in, 163; scientific and technical expertise of, 4, 7–8; Zionist movement in, 170–71, 177–78. *See also* Congress, U.S.; *specific departments and agencies*
University of California, Berkeley, 14, 24; agricultural training and, 57; Irrigation Department, 84; Mead's association with, 86, 89, 133, 161, 167, 171; Palestinian agricultural development and, 171, 172; Rural Institution Department, 86; scientific colonization and, 90
Upper Doab district (India), 31–32
Urbanization, 93, 150, 151; settlers' retreat from, 76
Utah, 52, 140
Utopianism, 72–73, 165; technological, 8–9, 16

Vaal River, 100, 110, 111, 126

Van Nuys, Calif., 89

Veblen, Thorstein, 182–83; *The Engineers and the Price System*, 12, 182

Vegetable farming, 92

Venezuela, 99

Victoria, Australia, 13, 14, 54, 67–87; California compared with, 66, 67–68, 70, 71, 85; California engineering ideals and, 19, 66, 180, 183, 185; closer settlement growth in, 82–87, 89, 155, 175–76, 180; land and irrigation policies and, 69, 94, 128; Mead's irrigation expertise and, 83–95; model irrigation colony of, 73, 74–87, 90, 94, 127, 166, 185; ninety irrigation trusts in, 73–74

Victoria (queen of Great Britain), 117, 180

Victorian Act of 1909, 90

Victorian Land Settlement Delegation, 85–86

*Victorian Settlers' Guide* (newspaper), 85

Victorian Water Act of 1905, 83

Victorian Water Supply Commission, 69–70

Victoria State Rivers and Water Supply Commission, 82, 83, 84, 85. *See also* Closer settlement program

Vierfontein, 111

Vincent, J. E. Matthew, 75, 76

Viticulture, 54, 78; Palestine and, 164, 168; South Africa and, 121, 122, 126; wine market and, 81, 82, 164

Vulcan Iron Works, 100


Waiakea homestead (Hawaii), 156

Waikapu Commons, 138

Wailuku Sugar Company, 137

Warburg, Otto, 168, 172

Warracknabeal, Australia, 75

*Water and Land* (journal), 49

Water Conservation Act of 1883 (Victoria), 69

"Water courts," 128

Water districts, 63

Water engineering. *See* Engineering; Hydraulic engineering; Irrigation

Water Law of 1959 (Israel), 178

Waterman, Robert, 49

Water policy, 11, 13, 14, 18, 19–20; Austra-

lian nationalization and, 69, 72, 83, 166; British India nationalization and, 30, 32–33, 34, 47, 73, 166; California landmark cases and, 49–50; California litigation and, 128; California's antiquated laws and, 27–28, 36, 40, 42, 47–49, 66, 72, 138; centralization and need for, 178; community ownership and, 51, 53, 56, 58, 178; dangers of private ownership and, 128; federal vs. states' rights approach to, 84; government vs. private monopoly ownership of, 25, 26, 34, 47–48; Hall water law writings and, 184–85; Hawaiian Islands and, 138; Israeli water law and, 178; Johannesburg and, 97, 110, 111; monopolies and, 21, 23, 25, 27, 40, 48, 139; mutual water companies and, 52, 53–54, 56, 61, 75; Palestine and, 178; public ownership advocacy and, 128; riparian principle-prior appropriation conflict and, 19, 22, 23, 27, 35, 42; San Joaquin project and, 21–26, 27; South African mining and, 110–11; South African proposed water districts and, 128–29; state-monitored private enterprise and, 48; state total control of, 73; U.S. Reclamation Service and, 59–60; Wyoming model rights laws and, 49, 83–84, 178. *See also* Irrigation; Prior appropriation doctrine; Riparian principle

Water Users' Association (Imperial Valley), 62

Waterworks Construction Act of 1886 (Victoria), 74

Weizmann, Chaim, 170, 171, 177

Wells Fargo, 22

Weribee project (Australia), 80

Wernher, Beit, and Company, 110, 111

Western Australia, 69

Western Europe, 4, 7, 31; racial hierarchy and, 6, 8, 15, 30

Western Sugar Company, 136, 149

Western Sugar Refinery, 134, 141

West Maui Mountain, 138

West Maui sugar plantation, 137

West Point (U.S. Military Academy), 24, 123

West Side Irrigation Commission (Calif.), 41

West Side Irrigation District (Calif.), 37

Whaling ships, 134–35

Wheat. *See* Grain production

Wheeler, Benjamin Ide, 86

Widney, Robert M., 56, 57

Williams, Gardner F., 100–101, 117

Wilson, Woodrow, 184

Wine market, 81, 82, 164

Wittfogel, Karl, 30–31

Witwatersrand (Rand), 97, 100, 101, 110, 111; Boer-English conflict and (*see* Anglo-Boer War); Chamber of Mines and, 115; Johannesburg and, 101; Reform Committee and, 116, 117; world's largest gold field in, 102–3, 110, 120. *See also* Hammond, John Hays; Mining engineering

Wolf, Eric, 8

Wolman, Leo, 176

Wool production, 81, 121

Works, John D., 86

World federation, 6

World's fairs, 1–8, 16, 56, 179

World's Social Progress Council, 6–7

World War I, 6, 16, 163, 167, 185; California veterans' land settlement program and, 92; Hoover relief work and, 184

World War II, 86

World Zionist Organization, 165, 170, 177, 178

Worster, Donald: *Rivers of Empire*, 31

Wright Act of 1887 (Calif.), 50, 53

Wycheproof, Australia, 75

Wyoming, 14; model water laws of, 49, 83–84, 178

Yale University, 118

Yeoman farmer ideal, 20, 26–27, 39, 126; Chaffey brothers' colonies and, 57; Hawaiian Islands and, 131, 133, 148, 154–55, 160; Mead's beliefs and, 87–88, 89, 90, 95, 150–51; romanticization of, 76, 165; threats to, 88; Victoria settlers and, 67. *See also* Rural life

Yuba City, Calif., 35, 44, 89

Yuba River, 18; mining debris and, 35, 43, 44, 46–47, 50

Yuma, Calif.: federal reclamation project, 60, 62, 63, 150

Yuma Indian Reservation, 61–62

Zionist Executive, 177

Zionist movement, 95, 160, 161, 162, 165–66, 170–71, 177–78; haphazard colonization and, 172, 175; idealism of, 165, 174, 185; Mead's endorsement of, 170

Zionist Organization, 165, 166, 167–68, 169, 170, 171

Zululand, 103

Zuni Indians, 30